TO MOVE THE WORLD

TO MOVE
THE WORLD

JFK'S QUEST

for

PEACE

Jeffrey D. Sachs

RANDOM HOUSE

NEW YORK

Published in the United States by Random House,
an imprint of The Random House Publishing Group,
a division of Random House, Inc., New York.

RANDOM HOUSE and colophon are registered
trademarks of Random House, Inc.

Library of Congress Cataloging-in-Publication Data
Sachs, Jeffrey.
To move the world : JFK's quest for peace / Jeffrey D. Sachs.
p.   cm.
Includes bibliographical references and index.
ISBN 978-0-8129-9492-6
eBook ISBN 978-0-8129-9493-3
1. Kennedy, John F. (John Fitzgerald), 1917–1963.   2. Kennedy, John F.
(John Fitzgerald), 1917–1963—Oratory.   3. United States—Politics and
government—1961–1963.   4. United States—Foreign relations—Soviet
Union.   5. Soviet Union—Foreign relations—United States.   6. World
politics—1955–1965.   I. Title.
E841.S16 2013
327.7304709'046—dc23
2013007914

Printed in the United States of America on acid-free paper

www.atrandom.com

2   4   6   8   9   7   5   3   1

First Edition

Book design by Christopher M. Zucker

For Sienna
We all cherish our grandchildren's futures

CONTENTS

Preface   *xi*

Chapter 1: THE QUEST FOR PEACE   3
Chapter 2: TO THE BRINK   26
Chapter 3: PRELUDE TO PEACE   39
Chapter 4: THE RHETORIC OF PEACE   50
Chapter 5: THE PEACE SPEECH   70
Chapter 6: THE CAMPAIGN FOR PEACE   91
Chapter 7: CONFIRMING THE TREATY   112
Chapter 8: THE HISTORIC MEANING OF
   KENNEDY'S PEACE INITIATIVE   137
Chapter 9: LET US TAKE OUR STAND   156

The Speeches:

AMERICAN UNIVERSITY COMMENCEMENT
   ADDRESS, JUNE 10, 1963   170
SPEECH TO THE IRISH DÁIL, JUNE 28, 1963   180
ADDRESS TO THE NATION ON THE PARTIAL
   NUCLEAR TEST BAN TREATY, JULY 26, 1963   190
SPEECH TO THE 18TH GENERAL ASSEMBLY
   OF THE UNITED NATIONS, SEPTEMBER 20, 1963 200

Acknowledgments   211
Bibliography   215
Notes   219
Index   231

PREFACE

"THERE'S ALWAYS SOME son-of-a-bitch who doesn't get the word," John F. Kennedy exclaimed in frustration. The president and his advisers were huddled at the height of the Cuban Missile Crisis in October 1962, with the United States and the Soviet Union on the brink of nuclear war. Kennedy had grounded U-2 spy-plane missions to avoid a provocation that might lead accidentally to a shooting war. Yet one Alaska-based U.S. Air Force pilot had not gotten the message. After taking off to collect air samples to check on Soviet nuclear testing, the pilot had become disoriented and inadvertently flown his plane into Soviet airspace. Soviet fighter jets scrambled to intercept the U-2, while due to the high alert status prompted by the crisis, the U.S. planes sent to escort it back to base were armed with nuclear warheads and had the authority to fire.[1] By dumb luck the world survived.[2]

Such was the slender thread of humanity's survival on those bleakest of days, the closest that the world has ever come to self-destruction. Looking back fifty years, it's hard to imagine, or even to believe, that humanity nearly squandered everything over issues and causes that don't even exist today. One of the two super-powers no longer survives. The cause of global communism is defunct, a failed idea that was abandoned by its own protagonists more than two decades ago. The politics of Cuba and Berlin, two of the most intractable conflict points of the Cold War, hardly seem matters on which human survival should turn.

We are tempted to say "Never mind" about a struggle that makes little sense in today's context. Yet we mustn't turn our back on that history. We must understand how two superpowers not only came to the brink of global annihilation, but built thousands of nuclear warheads, including single warheads that had vastly more destructive power than all the bombs dropped in World War II combined. We must also remember that the superpower conflict was not really "cold" at all, as it resulted in countless "proxy wars" that claimed millions of lives across several conti-nents. We are still living in the world created by the Cold War, and we are still living under the shadow of thousands of nuclear war-heads, even if their numbers are down and the hair-trigger rules for their deployment are gone. We also need to understand the Cold War because the human instincts and political institutions that created it still shape our country's approach to the world and the strategies of other countries. Many of today's unnecessary conflicts are directly descended from the dynamics and outcomes of the Cold War.

There is a more positive and dynamic reason to understand those times as well. The great turning point of the Cold War, the stepping back from the nuclear abyss, was an act of political grace and courage, led by President John F. Kennedy and his Soviet counterpart, Communist Party chairman Nikita Khrushchev. These two men, in their distinct ways, both said "Enough." They came to the realization that the world could not afford to lurch from crisis to crisis, with one son of a bitch or another failing to get the message and thereby plunging the world into horror and darkness. Kennedy and Khrushchev knew too well that their own colleagues—thoughtlessly, ruthlessly, stupidly, or naïvely—might be contributors to such a world-shattering blunder. We must understand how humanity was saved from the accidents, miscalculations, bravado, and supposedly sophisticated strategic thinking that almost ended it all.

This book recalls John F. Kennedy's *annus mirabilis,* from October 1962 to September 1963, when he and Khrushchev saved the world, and left a legacy, a blueprint, and an inspiration for those who would follow. Kennedy had come to office in January 1961 inexperienced, the youngest elected president in American history. Like all national leaders of the day, he was a Cold Warrior himself, determined to preserve American liberty in the face of a perceived threat of global communism. Yet he was also determined from the first day of his administration to find a path to peace. That path was unclear, and both Kennedy and Khrushchev would stumble badly along the way, from the Bay of Pigs to the Cuban Missile Crisis.

Despite these errors, missteps, and near disasters, and because of the lessons learned from them, Kennedy and Khrushchev found a path back from the brink and toward the peaceful resolution of the Cold War. Kennedy campaigned for peace on three fronts: with Khrushchev, both an adversary and partner; with the U.S. allies, who were never simple and often divided on key issues; and with the U.S. political system, which was deeply entrenched in the Cold War and not easily moved toward peace. Kennedy's peace campaign found its greatest eloquence during the summer of 1963, leaving us a legacy of words and deeds of historic proportion.

Fifty years on is a fitting and appropriate time to recall these events and to seek to learn from them. Yet this project jelled in my mind several years ago, when I fell in love with Kennedy's great proclamation on peace, his "Strategy of Peace" address given as the American University commencement address in June 1963. This "peace speech," although admired by some as one of Kennedy's finest, has never achieved the fame of his inaugural address or the great speech on civil rights that he delivered just one day after the Peace Speech. I had not really known the Peace Speech until I came across it a few years ago while working on issues of global poverty. The speech moved me deeply, not only for its eloquence and content, but also for its relevance to today's global challenges. For in it Kennedy tells us about transforming our deepest aspirations—in this case for peace—into practical realities. He almost presents a method, a dream-and-do combination that soars with high vision and yet walks on earth with practical results.

I included a discussion of these remarkable features of the Peace Speech in the Reith Lectures that I gave for the BBC in 2007.[3] The third of these lectures took place at Columbia University, my home institution, and the event was most special because of the man sitting front and center in the first row: Ted Sorensen, John Kennedy's intellectual alter ego, counselor, and gifted speechwriter, the man who had not only drafted the Peace Speech but had worked intimately with Kennedy for a decade on the concepts and dreams articulated in it. Sorensen was much more than a draftsman. He was a moral force and intellectual partner of John Kennedy, and the Peace Speech is a product of and tribute to the Kennedy-Sorensen duo in the deepest sense.

When I finished the address, I was humbled by Sorensen's telling me that I had "gotten it." I was moved and excited to hear from him that the speech was his favorite of all Kennedy's speeches, which comported with my own understanding of its unique importance in Kennedy's foreign policy and in modern world history. Here indeed was a case where words, magnificently crafted and carefully thought through, had made a difference.

At the time I worked closely with Sorensen's remarkable wife, Gillian, who was a senior adviser of United Nations Secretary-General Kofi Annan. My wife, Sonia, and I had wonderful opportunities to discuss current events with Ted and Gillian Sorensen, our neighbors. On those occasions, I would discuss the Peace

Speech with Sorensen, which deepened both my understanding of its context and my profound admiration for it. I decided at that point that I would write about it in a book, and asked for Sorensen's advice and help. He greatly appreciated the project, and was eager to work together to document the speech, its history, and its impact. Sadly, Ted Sorensen was struck down by a devastating stroke not long after. A man of remarkable talents, ethics, experience, and wit was gone. My grand hope to work with him on a book about the Peace Speech was lost. How many times during the writing of this book did I wish that I could have given Ted a call to get guidance and reflections!

In reviewing the history and context of the Peace Speech, my esteem for it and for Kennedy has only grown. I have come to believe that Kennedy's quest for peace is not only the greatest achievement of his presidency, but also one of the greatest acts of world leadership in the modern era. I certainly do not pretend that the speech alone changed history, or that it marked the end of the Cold War. Nor do I want to leave the impression that peace was the work of one man, much less one speech. It was Kennedy himself who declared that there is no simple, single key to peace, that "genuine peace must be the product of many nations, the sum of many acts. It must be dynamic, not static, changing to meet the challenge of each new generation. For peace is a process—a way of solving problems."

The speech remains important for us today not only for what it did, but for what it tells us about the process of making peace and social reform more generally. We have new challenges in our generation, the most important of which is the challenge of sustainable development: learning to live together on a crowded planet, in harmony not only with more than seven billion others but with a physical earth under dire assault from a burgeoning world economy. I find the Peace Speech a wonderful help in thinking about that challenge too. Kennedy noted that the core of our common humanity is that "we all inhabit this small planet. We all breathe

the same air." That air is increasingly threatened. In Kennedy's day the dire threat to the air was nuclear fallout. In our time it is greenhouse gases. But in both cases the underlying truth is exactly the same: we need to make earth a fitting home for all of humanity.

Words can move us to great deeds. In Kennedy's case, the words inspired both Americans and Soviets to take the risk for peace by adopting a treaty on nuclear testing, which had proved elusive till then and which was opposed strenuously by hardliners on both sides. Kennedy's words shaped a common understanding of what was possible for mutual benefit, helping to break the hammerlock of fear and loathing. We will look at those words closely: their beauty, their provenance, and their revelation of Kennedy's own growth as a leader. I've included it and three other key speeches at the end of the book for readers to savor in full.

Let us therefore learn and marvel at how Kennedy helped humanity to take one more step on the path of survival and human achievement.

TO MOVE THE WORLD

Chapter 1.
———————

THE QUEST FOR PEACE

WHEN JOHN F. KENNEDY came to office in January 1961, the world lived in peril of a nuclear war between the two superpowers. The Cold War confrontation between the United States and the Soviet Union would eventually consume trillions of dollars and millions of lives in wars fought around the world. At times, humanity seemed to be "gripped by forces we cannot control," a pessimistic view that Kennedy noted and strenuously argued against in his Peace Speech. And yet the power of those disruptive forces at times was indeed nearly overwhelming, causing events to spin beyond the control even of presidents, Communist Party chairmen, and the countries they led.

The Cold War was in every sense a stepchild of the two world wars. Those wars created the structures of geopolitics, military might, and, perhaps most important of all, the psychological mindsets that determined the course of the Cold War. John Kennedy's peace strategy would emerge from his intimate understanding of the dynamics that had driven the two wars. The first war he knew as a voracious student of history, especially the history as written by Winston Churchill. The second war he knew firsthand. The years between 1938 and 1945 were a deeply formative period of his adult life—as a student in prewar London while his father was U.S. ambassador to the United Kingdom; as a young author grappling with the question of why England had failed for so long to confront Hitler; as a patrol boat captain in the Pacific, where his vessel PT-109 was sunk by a Japanese destroyer; and as part of a grieving family when his elder brother was lost in a daring bombing mission over Germany.[1]

The overwhelming question facing the world, and facing Kennedy during his presidency, was how to prevent a third world war. The factors that had caused the two wars—geopolitics, arms races, blunders, bluster, miscalculations, fears, and opportunism—continued to operate and to threaten a new conflagration. Yet the context was also fundamentally new and more threatening. The nuclear bombs dropped on Hiroshima and Nagasaki in 1945 to conclude World War II had ushered in the nuclear age, and had made the stakes incalculably higher. A thermonuclear bomb could now carry far more explosive force than all of the bombs of the Second World War.

Kennedy's worldview on these issues was shaped above all by the influence and model of Winston Churchill, England's great author-politician-warrior-statesman, whose masterly history of the first war, *The World Crisis*, described a tragic era of war through miscalculation;[2] whose warnings about Hitler in the 1930s had gone unheeded until almost too late; whose leadership as prime minister between 1940 and 1945 enabled the United Kingdom to survive and eventually triumph over Hitler; whose warnings in 1946, just after World War II ended, alerted the West to the rising threat of Soviet power; and whose calls during the 1940s and 1950s for a negotiated settlement with the Soviet Union did much to influence Kennedy's peace strategy as president.[3]

Kennedy's lifelong fascination with, learning from, and urge to emulate Winston Churchill has been recounted by many biographers.

The greatest problem facing Kennedy (and indeed the world) in drawing lessons from the two world wars was that the lessons were highly complex, subtle, and even seemingly contradictory. World War I seemed to be a lesson about self-fulfilling crises, where the fear of war itself led to an arms race, while the arms race in turn led to a world primed for war. These lessons seemed to call for restraint in the arms race and avoidance of a self-fulfilling rush to war, and so even as Hitler rearmed Germany in the 1930s, in contravention of the Treaty of Versailles that had ended World War I, Britain avoided provocations that could spiral out of control. Most famously, Prime Minister Neville Chamberlain argued that it would be better to accede to German demands on border adjustments with Czechoslovakia, and so "appeased" Hitler in the name of peace at the Munich conference in 1938, a disastrous mistake that fueled Hitler's drive to war.[4]

If World War I seemed to argue against arms races and self-fulfilling prophecies of war, the lead-up to World War II, by contrast, seemed to argue for meeting strength with strength, and avoiding the temptation of "appeasement." For Kennedy, the debate over appeasement was more than intellectual; it was intensely personal. John Kennedy watched closely as his father, Joe, strongly defended appeasement, indeed declaring that Chamberlain had no choice when he acceded to Hitler's outrageous demands at the 1938 Munich conference, as Hitler would have defeated the United Kingdom in battle. When war finally broke out, the proponents of appeasement were humiliated and Joe Kennedy's vast political ambitions were destroyed.[5] The younger Kennedy would soon implicitly come down on Churchill's side, writing in his first book, *Why England Slept,* that Britain had dangerously delayed rearming under the illusion that appeasing Hitler would keep it safe and out of war.[6]

As president, Kennedy would battle with these powerful and conflicting dynamics. Should he restrain the arms race in order to avoid a self-feeding race to war with the Soviet Union? Or should he strengthen U.S. arms in order to negotiate from strength? Should he make concessions to the Soviet premier Nikita Khrushchev to acknowledge Soviet interests? Or should he hold the line to avoid the appearance and reality of appeasement? Kennedy would remain a student of history, and of Churchill, whom he most admired, trying to apply the complex lessons of the past to the urgent challenges of the present.

The Nuclear Arms Race

The problems of distrust between the Soviet Union and the United States were profound, pervasive, and persistent, and that distrust spurred the arms race. The two sides were of course rivals and competitors. And each side lied to the other, repeatedly and persistently. These were not grounds for easy trust. Nor was the historical context. Just a few years earlier, Hitler had cheated relentlessly, thereby winning significant concessions. Chamberlain's appeasement of Hitler at Munich hung over the Cold War era: Don't trust the other side. Better to arm to the teeth.

Even though there were enormous gains to be had by both the United States and the Soviet Union if they could agree on the postwar order in Europe, politicians on both sides found it nearly impossible to take any steps that required trust. If they did, they opened themselves up to extraordinarily harsh attacks by hardliners on their own side who denied that the other side would abide by any agreements. A U.S. politician who urged agreement with the Soviet Union risked immediate subjection to the cries of

"Munich" and "appeasement," powerful political charges and ones Kennedy was especially eager to avoid.

The two sides were trapped by two closely related problems: the prisoner's dilemma and the security dilemma. The prisoner's dilemma holds that in the absence of long-term trust or binding agreements, the logic of inter-state rivalry will push both sides to arm. Should the United States arm or disarm? If the Soviet Union arms, the United States has no choice but to arm as well in order to avoid being the weaker side. If the Soviet Union disarms, then the United States gains military and political advantage by arming while the Soviet Union is weak. Therefore, arming is a "dominant" strategy: the best move no matter what the other side does. Since the logic is the same for the other side, both sides end up continually increasing their arms, even though a binding agreement to disarm would be mutually beneficial.[7]

The security dilemma, propounded by Robert Jervis, a leading political theorist, is a corollary of the prisoner's dilemma.[8] The security dilemma holds that a *defensive* action by one side will often be viewed by the other side as an offensive action. Thus, if the United States builds its nuclear arsenal to stave off a Soviet conventional land invasion of Europe, the Soviet Union will view the U.S. nuclear buildup as preparation for a nuclear first strike against the Soviet Union rather than as a defensive measure. And if the Soviet Union tries to catch up with the U.S. nuclear arsenal, that will be viewed as an offensive action by the United States. U.S. hardliners would argue that the Soviet Union is trying to neutralize the U.S. nuclear deterrent so that the Soviet Union can launch a conventional attack.

As a result of the absence of trust, and the harsh logic of both the prisoner's dilemma and the security dilemma, both sides continued to amass nuclear weapons to the point of massive overkill. And as the arsenals continued to expand, each side feared that the other was actually building up for a surprise first-strike attack.

The United States indeed contemplated launching a preventive nuclear war, worried that it would be unable to defend itself in the future. Jervis recalled the words of the German statesman Otto von Bismarck, who called a preventive war "committing suicide from fear of death."[9]

The nuclear arms race accelerated as the United States and the Soviet Union expanded their arsenals, and as the United Kingdom and France became nuclear powers (in 1952 and 1960, respectively) with their own independent arsenals. By 1960, the United States had nuclear warheads positioned in several countries around the world.[10] The Soviet Union felt itself very much surrounded indeed, and increasingly unsure of whether these U.S. nuclear weapons were really under U.S. control.

Of course it wasn't just the international situation that prompted the arms buildup on each side. It was also domestic politics. The military-industrial complex gained power within each government as time went on. In the United States, each branch of the military demanded its own nuclear arsenal, so that competition among the U.S. Army, Air Force, and Navy also drove up military budgets and the numbers of nuclear warheads and delivery systems. The same was true on the Soviet side, where there was far less constraint than in the United States on the political power of the military-industrial complex.

To the Brink

When Kennedy assumed office, he took to heart Churchill's belief that political leaders must work actively to solve vexing international problems. He was intent on pursuing arms control, but was also a staunch Cold Warrior, partly out of conviction and partly out of political expediency, in order to protect himself from powerful hardline anti-communists. Kennedy believed that he could

President Dwight D. Eisenhower greets President-elect John F. Kennedy (December 6, 1960).

untangle the dangerous conflicts with the Soviet Union. And, as Churchill urged, Kennedy would aim to solve these problems from a position of U.S. military strength and without relinquishing vital Western interests.

The tough and conciliatory sides of Kennedy's negotiating strategy were mutually reinforcing. Churchill had long emphasized the essential role of negotiating with one's adversary: "To jaw-jaw," he said, "is always better than to war-war."[11] Churchill had called negotiation through strength his "double-barreled strategy," and famously declared, "I do not hold that we should rearm in order to fight. I hold that we should rearm in order to parley."[12] In 1938, it had not been just a weakness of political will but also one of military preparedness that had led Chamberlain to appease Hitler at Munich.

Kennedy would refer to Churchill's double-barreled approach in his campaign address in Seattle in September 1960:

> It is an unfortunate fact that we can secure peace only
> by preparing for war. Winston Churchill said in 1949,
> "We arm to parley." We can convince Mr. Khrushchev
> to bargain seriously at the conference table if he respects
> our strength.[13]

Kennedy had no doubt either of the enormous potential gains
of cooperation with the Soviet Union, or of the grave risks if the
United States cooperated (for instance, through arms control)
while the Soviet Union reneged on its side of the deal. Cheating
by the Soviet Union would threaten not only U.S. security, but
also Kennedy's hold on power domestically. Kennedy would repeatedly urge cooperation but remain alert that any move toward
cooperation, however modest, could trigger political charges from
the right that he was an appeaser.

The burdens on Kennedy were greater as a Democrat, since Republicans regularly assailed the Democratic Party for being "soft
on communism." Kennedy therefore aimed to assure all sides—
the U.S. public, America's allies, and of course the Soviet Union—
that he would vigorously resist Soviet aggression and defend
Western interests while he sought greater cooperation with the
Soviet Union. He would aim, at the core, to pursue a tit-for-tat
strategy (a way to break out of the prisoner's dilemma by recipro-
cating cooperation from the other side), promising to join the
Soviet Union in arms control, but also declaring repeatedly his
readiness to revert to an arms race if the Soviet Union did not
keep its promises. The tit-for-tat strategy of incremental cooperation was mapped out a year after Kennedy came to office by one of
America's leading sociologists, Amitai Etzioni, whose remarkable
book *The Hard Way to Peace* spelled out a psychological approach
to forging peace.[14] Etzioni believed that confidence building was
crucial, since in his view psychological rather than political or
military factors were the decisive drivers of the Cold War. He propounded a notion of "psychological gradualism" to reduce fear,

build trust, and initiate a phased process of reciprocated concessions. Eventually suspicion and fear would be "reduced to a level
where fruitful negotiations are possible."[15] In many ways, Kennedy's peace initiative in 1963 would pursue this approach.

Kennedy first signaled both aspects of his approach in his
inaugural address on January 20, 1961.[16] First came his robust, full-
throated commitment to the defense of liberty:

> Let every nation know, whether it wishes us well or ill,
> that we shall pay any price, bear any burden, meet any
> hardship, support any friend, oppose any foe, in order
> to assure the survival and the success of liberty.

But equally stirring was his commitment to pursue the mutual
gains of cooperation:

> So let us begin anew—remembering on both sides that
> civility is not a sign of weakness, and sincerity is always
> subject to proof. Let us never negotiate out of fear. But
> let us never fear to negotiate . . .
>
> Let both sides, for the first time, formulate serious
> and precise proposals for the inspection and control of
> arms—and bring the absolute power to destroy other
> nations under the absolute control of all nations.
>
> Let both sides seek to invoke the wonders of science
> instead of its terrors. Together let us explore the stars,
> conquer the deserts, eradicate disease, tap the ocean
> depths, and encourage the arts and commerce.

Kennedy's emphasis on "precise proposals" was not incidental.
Following Churchill once again, Kennedy believed that miscalculation with the Soviet Union would best be avoided through clear,
detailed, and principled negotiating positions. Yet here too the
ideal and the practical would collide. Negotiations are filled with

feints, bluffs, and intermediate positions, and these inevitably raise the risk of miscalculation.

Kennedy was sincere in his inaugural address when he said, "So let us begin anew." As a senator and as a presidential candidate he himself had helped to spur the arms race by opportunistically hammering away at Dwight D. Eisenhower's administration for allowing a "missile gap" to emerge, claiming in 1958 that there was every indication "that by 1960 the United States will have lost . . . its superiority in nuclear striking power."[17] In fact, the true missile gap, contrary to Kennedy's claim, stood greatly in America's favor. While the definitive knowledge of the Soviet Union's very limited ballistic missile capacity was a closely held secret of the Eisenhower administration, based on secret U-2 spy plane flights, Senator Kennedy probably knew that he was exaggerating Soviet capabilities. As the presidential candidate running against hardline vice president Richard Nixon, Kennedy was especially keen to project a tough-minded foreign policy stance and avoid the charge of being soft on communism typically levied against Democrats. No matter what Kennedy may have believed as a candidate, he learned early in his presidency that there was no missile gap in the Soviets' advantage.

Khrushchev would probably have understood the political reasons for Kennedy's missile-gap rhetoric, and might even have benefited in a way from Kennedy's exaggerated portrayal of Soviet power. Indeed, when Kennedy chose in October 1961 to reveal to the public the relative weakness of the Soviet nuclear force, Khrushchev was deeply aggrieved. Still, in the early days of the new administration, Khrushchev was intent on determining whether Kennedy was in fact a diehard Cold Warrior, like Eisenhower's influential Secretary of State John Foster Dulles, or a potentially cooperative counterpart in peaceful coexistence. Early steps by Kennedy could therefore help to chart the course toward better relations.

In addition to their public speeches and the interaction of their diplomats, the two leaders would soon learn much more about each other in another way. Beginning with Khrushchev's letter of congratulations to the newly elected Kennedy on November 9, 1960, the two men engaged until Kennedy's death in a back-channel personal correspondence of more than a hundred letters. Both pledged that the letters would be held confidentially and never leaked for propaganda purposes. "For my part," Kennedy wrote, "the contents and even the existence of our letters will be known only to the Secretary of State and a few other of my closest associates in the government."[18] Both leaders seem to have honored their pledges of confidentiality. The result is a most extraordinary exchange that together with other events offers critical insights into the thinking of both men, the issues that worried them, and their strategies for peace.

Kennedy's Opening Provocations

Unfortunately for the prospects of U.S.-Soviet cooperation, and contrary to the approach of incremental cooperation, Kennedy came out swinging. He did this in three provocative ways. First, despite the fact that the United States was far ahead in nuclear weapons, Kennedy ordered a major military buildup of both nuclear and conventional arms. The total number of U.S. nuclear warheads would soar from 20,000 in 1960 to 29,000 in 1963, at a time when the Soviet Union had a small fraction of that number (1,600 in 1960 and 4,200 in 1963).[19] Conventional forces were also greatly augmented, as Kennedy adopted a new model of "flexible response." He was highly critical of Eisenhower's nuclear policy of "massive retaliation" to meet Soviet threats, which purportedly relied on U.S. nuclear weapons to deter Soviet provocations. Kennedy wanted more non-nuclear options.[20]

Second, Kennedy approved a CIA plan for an invasion of Cuba, which would become the biggest blunder of his presidency. Third,

he went ahead with a confrontational move that had been approved by his predecessor, Dwight Eisenhower. In 1958, Eisenhower had decided to strengthen the U.S. nuclear arsenal by
posting intermediate-range nuclear missiles under U.S. control in
Italy and Turkey. The placements of these Jupiter missiles were
implemented in June 1960 in Italy and October 1961 (the first year
of the Kennedy administration) in Turkey.[21] The Soviet Union now
faced the threat not only of America's strategic bombers but also
of nearby missiles that could reach the Soviet Union in minutes.
This was a new and terrifying prospect that tipped the psychological and strategic balance toward the United States and constituted a major motivation for Khrushchev's later attempt to put
similar missiles into Cuba.

Kennedy sought to negotiate peace through strength, but these
early moves were more than a mere show of strength: they were
a ratcheting up of the Cold War. Here the contradictory lessons
of World War I and World War II were starkly revealed. Kennedy
was profoundly concerned about a war starting through miscalculation, as had World War I, but was equally if not more
concerned with being perceived as weak if he failed to project military strength and firmness. Yet by taking steps to build U.S.
military strength and increase the number of U.S. military options, he inadvertently exacerbated the risks of terrible miscalculation, very much a case of Jervis's security dilemma in operation.

At the time that Kennedy assumed the presidency, the CIA was
already in high gear to topple Cuba's new left-wing government,
which had begun to confiscate U.S. assets and sidle up to the
Soviet Union. America had long backed Cuba's corrupt dictator,
Fulgencio Batista, who offered privileges and protection to American investors in the nearby island, only ninety miles from Florida.
During the 1950s, the young lawyer Fidel Castro led a guerrilla
insurgency against Batista, finally succeeding in prompting the
dictator to flee on January 1, 1959.[22] No sooner had Castro con-
solidated his control over Cuba than Eisenhower and the CIA di-

rector, Allen Dulles (the brother of Secretary of State John Foster Dulles), began to plot a coup to bring him down. Castro was not yet a hardcore Soviet ally, though a partial U.S. trade embargo, initiated by the Eisenhower administration in 1960, was pushing Cuba in that direction. U.S. actions in 1961 would soon lead Cuba fully into the Soviet camp. The CIA briefed Kennedy early in his term about plans for a CIA-backed invasion of Cuba that would be carried out by Cuban expatriates. The planning was far advanced, as the CIA had prepared it in the final year of the Eisenhower administration.

From the start, Kennedy was deeply concerned not only about the invasion's chances of success, but also about its implications for U.S.-Soviet relations. This would be among the first major moves in his new strategic game with Khrushchev, and it would be far from a cooperative one. Kennedy feared specifically that any action against Castro might prompt Soviet retaliation in Berlin, the hotspot of the Cold War. This linkage was probably exaggerated—the Soviets did not yet see Castro as vital—but it played a role in Kennedy's thinking. The planned invasion was also one of Kennedy's first foreign policy decisions. He did not yet have the confidence to disregard the CIA and the military.

Kennedy tried to have it both ways, and ended up in a disastrous muddle. He gave the green light to the CIA-based Cuban invasion in April 1961, but he wanted "deniability" of U.S. involvement and so withheld key military backing, such as air support, that was vital to any chance of military success. The hope of deniability was foolish; the U.S. role was obvious. Kennedy's prevarication guaranteed a complete failure of the attack (a failure that was most likely in any event), followed by harsh international criticism of the United States. The operation was too small for success, but too large for deniability. The expatriates who landed at the Bay of Pigs were quickly killed or captured, and the entire episode ended ignominiously.[23]

In the postmortem, Kennedy met with Eisenhower to discuss

the botched operation, and Eisenhower asked Kennedy why he had denied air cover for the invasion. The historian Michael Beschloss described the painful interchange:

> Kennedy said, "Well, my advice was that we must try to keep our hands from showing in the affair." Eisenhower was aghast: "How could you expect the world to believe we had nothing to do with it? Where did these people get the ships to go from Central America to Cuba? Where did they get the weapons? . . . I believe there is only one thing to do when you go into this kind of thing: it must be a success."[24]

By itself, the CIA operation was foolish, naïve, and incompetently designed and managed. This was par for the course for the CIA, which had bungled one operation after another in many parts of the world. The Bay of Pigs fiasco also confirmed Kennedy's deep mistrust of the military, which had begun with his experience in World War II. Kennedy told reporter and friend Ben Bradlee, "The first advice I'm going to give my successor is to watch the generals and to avoid feeling that just because they were military men their opinions on military matters were worth a damn."[25]

It also contributed to a cascading set of errors. Forced to appear tough and decisive after the very public failure, Kennedy quickly called for increased military spending and harsher measures to destabilize the Castro regime.[26] These included a series of harebrained attempts to assassinate Castro, part of the larger anti-Castro strategy the CIA dubbed "Operation Mongoose."[27] As a result, both Castro and Khrushchev came to believe that Kennedy's next gambit would be a full-fledged invasion of Cuba. That expectation contributed to the Cuban Missile Crisis the following year.

The Kennedy-Khrushchev private letters offer a remarkable interchange regarding the Bay of Pigs. The letters before the Bay of Pigs are highly congenial, the opening moves of confidence building under the tit-for-tat strategy. Khrushchev writes on November 9, 1960, that he hopes Kennedy's election will mean that "our countries would again follow the line along which they were developing in Franklin Roosevelt's time," when the two countries were allies.[28] He holds out the prospect of important agreements: "we are ready, for our part, to continue efforts to solve such a pressing problem as disarmament, to settle the German issue through the earliest conclusion of a peace treaty and to reach agreement on other questions."[29] Kennedy responds that "a just and lasting peace will remain a fundamental goal of this nation and a major task of its President."[30] They begin to arrange an early summit meeting.

Yet the tone cracks in Khrushchev's letter of April 18, two days after the Cuban invasion:

> Mr. President, I send you this message in an hour of alarm, fraught with danger for the peace of the whole world. Armed aggression has begun against Cuba. It is a secret to no one that the armed bands invading this country were trained, equipped, and armed in the United States of America.[31]

"I approach you, Mr. President," Khrushchev writes, "with an urgent call to put an end to aggression against the Republic of Cuba."

Kennedy's answer the same day is dreadfully maladroit. "You are under a serious misapprehension in regard to the events in Cuba," he replies. "I have previously stated, and I repeat now, that the United States intends no military intervention in Cuba."[32] This patently false denial brings a powerful rebuke from Khrushchev four days later:

I have received your reply of April 18. You write that the United States intends no military intervention in Cuba. But numerous facts known to the whole world—and to the Government of the United States, of course, better than to any one else—speak differently. Despite all assurances to the contrary, it has now been proved beyond doubt that it was precisely the United States which prepared the intervention, financed its arming and transported the gangs of mercenaries that invaded the territory of Cuba.[33]

Kennedy's denial marked the second notable U.S. presidential lie to Khrushchev in less than a year. The previous summer, the CIA had pressed Eisenhower to permit another round of U-2 spy flights over the Soviet Union. He reluctantly agreed.* When a U-2 plane was shot down, the CIA and Eisenhower assumed that the pilot, Francis Gary Powers, had been killed and the plane destroyed in the crash. They therefore publicly lied about the mission, claiming that a U.S. weather-research plane had lost its way and crashed in Soviet airspace. Khrushchev then revealed Eisenhower's lie by producing not only the U-2 wreckage, but the live pilot as well. U.S. perfidy was exposed, and Eisenhower was forced to take responsibility. Yet this was not a simple public relations victory for Khrushchev. It was a bitter setback for Khrushchev's concept of peaceful coexistence. It also undercut his domestic credibility, as he had initially defended Eisenhower as not responsible for the U-2 flight, and the exposure of Eisenhower's lie seemed to give credence to the Soviet hardliners who argued that the United States could not be trusted. Khrushchev soon enough would demonstrate his capacity to lie about weighty matters as well; the Cold War was not a game played by saints. Yet the back-to-back prevarications by Eisenhower and Kennedy surely em-

* The U-2 flights were seen even by the United States as serious provocations to the USSR, and violations of international law.

boldened Khrushchev in his own future dissembling regarding nuclear testing and Cuba.

In a follow-up letter to Kennedy (May 16), Khrushchev acknowledges that "a certain open falling out has taken place in the relations between our countries."[34] Yet he set the ground for an upcoming June meeting with Kennedy in Vienna by emphasizing the key point he sees: the importance of a U.S.-Soviet settlement on Germany. The fate of postwar Germany had proved a constant source of tension between the two superpowers since the dawn of the Cold War, and never more so than when Kennedy took office. It is therefore crucial to revisit some of this history.

At the July 1945 Potsdam conference at the end of World War II, it was agreed that a council of the four major Allied powers (the United States, the United Kingdom, France, and the Soviet Union) would administer postwar Germany, but the "how" was left unspecified.[35] In the short term, the four powers accepted that Germany would be divided into four occupation zones and that each occupying power would manage its own zone until longer-term arrangements for a unified Germany could be agreed upon. The capital city of Berlin, though falling within the Soviet zone, was also divided among the four powers. But the longer-term arrangements for Germany were never agreed upon, and relations between the West and the Soviet Union quickly deteriorated once there was no common enemy of Hitler to unite them.[36] The United States, France, and the United Kingdom soon amalgamated their occupation zones into a single entity, which in 1949 became the Federal Republic of Germany (West Germany). The Soviet zone became the German Democratic Republic (East Germany). The failure to agree on the fate of a unified Germany sowed the seeds for many of the Cold War conflicts that followed.

Herein lay the basic dilemma. The Soviet Union—which had lost more than twenty million soldiers and civilians in World War II—feared a German resurgence, and thereby asserted harsh control over the Soviet occupation zone of Germany. Not only that, but Joseph Stalin, the brutal leader of the Soviet Union since the mid-1920s, ruthlessly created satellite states in Eastern Europe (in Yugoslavia, Czechoslovakia, Poland, and elsewhere) under single-party communist rule, thereby establishing a controlling corridor from Russia all the way to the heart of Germany. To the Western powers, these actions seemed to suggest a Soviet plan to dominate all of Europe, if not the world. While U.S. leaders had initially favored deindustrializing Germany at the end of the war to prevent it from again becoming an economic and military power,[37] they quickly changed their minds in the face of the perceived Soviet aggression. They instead decided to reindustrialize the western part of Germany as rapidly as possible as a bulwark against the Soviet Union and to strengthen Western Europe generally.[38] The Soviet Union viewed the hardening of U.S. positions as a violation of agreements intended to prevent a long-term resurgence of German power.

It's not hard to see where this led. As West Germany got stronger, Soviet anxieties rose. As the Soviet Union's anxieties rose, it became more belligerent in response, and the West then became even more determined to rebuild West Germany to resist Soviet domination. This explosive dynamic continued after Kennedy's assumption of office.

Making things even more difficult for Kennedy was the fact that Eisenhower had proved neither very attentive nor interested in the Soviet concerns vis-à-vis Germany. Eisenhower stood strongly in favor of Germany's economic and military recovery, in part because he wanted Western Europe to defend itself so that U.S. troops could return home.[39] Eisenhower liked neither the cost of stationing U.S. troops in Europe nor the long-term political commitment. Eisenhower believed that U.S. commitments should be tapered down as soon as Europe could take up the burdens of its own defense.[40] And if that even meant a Western Europe with nuclear weapons, he was generally for it.

In the final years of the Eisenhower administration, in response to the European desire for nuclear weapons, the United States and NATO—the military alliance of Western powers, formed in 1955—began floating the idea of "nuclear sharing" among the NATO countries, perhaps through a nuclear Multilateral Force (MLF). The United States saw the MLF as a way to share nuclear weapons with allies and give Europe its own deterrent without having to give any one nation full control and the power to launch a nuclear weapon unilaterally.[41] This prospect, particularly of West German nuclear access, was a crucial factor in the dramatic heating up of U.S.-Soviet tensions in the lead-up to Kennedy's presidency. This point is made forcefully by the historian Marc Trachtenberg in *A Constructed Peace* (1999), one of the most incisive and important historical analyses of this phase of the Cold War.

Kennedy did not yet have a strategy for Germany, and indeed was pressed hard by the West German leader Konrad Adenauer *not* to have one. From Adenauer's point of view, any thaw in relations between the United States and the Soviet Union would likely come at West Germany's expense. But Kennedy would listen to and learn from Khrushchev's repeated and heated concerns over Germany, especially concerning Germany's acquisition of nuclear weapons. Kennedy would eventually break with Adenauer on that question, thereby paving the way to closer cooperation with the Soviet Union. But that breakthrough was still two years in the future, and dire risks were strewn on the path to success.

Berlin Redux

Following the Bay of Pigs debacle, it was now Khrushchev's turn to misstep badly. With Kennedy on his back foot, Khrushchev believed that he was now in a position to pressure Kennedy into an agreement on Germany, which was Khrushchev's primary foreign policy concern. Kennedy's primary interest was in discussing a nuclear test ban treaty, which he felt was essential to slowing the

Soviet premier Nikita Khrushchev with President Kennedy at the start of the
Vienna Summit (June 3, 1961).

arms race and nuclear proliferation, one of Kennedy's most
pressing concerns. The idea of a test ban treaty had been discussed for several years, but the two sides were never able to get
past their disagreements on how such an agreement would be
enforced.

At the Vienna Summit of June 1961, Khrushchev ensured that
Germany and Berlin, rather than a test ban treaty, dominated
the discussion. Khrushchev told Kennedy that the Soviet Union
would recognize the German Democratic Republic (East Germany) and end Western access rights to West Berlin by the end of
1961. America and its allies viewed access to West Berlin as a vital
interest of the Western alliance, and so Khrushchev's threat was an
enormous provocation. If war came from this, it would come, said
Khrushchev. He said that Berlin was a "running sore," and that
disarmament was "impossible as long as the Berlin problem ex-

isted."[42] Kennedy responded, "Then, Mr. Chairman, there will be a war. It will be a cold, long winter."[43]

In fact, the difficult interchange brought the two leaders a little closer to a solution, though they certainly could not see it at the time. There were actually three German issues on the table, all intricately intertwined. The first was the Soviet desire for a peace treaty with Germany that would somehow protect the Soviet Union from future German aggression. The second was West Berlin, which Khrushchev considered to be "this thorn, this ulcer" within East Germany.[44] Khrushchev viewed West Berlin as a staging post for Western spying and aggression against the Soviet Union, "a NATO beachhead and military base against us inside the GDR."[45] He wanted the Western troops out. The third issue involved the question of German rearmament, and especially German access to U.S. nuclear weapons. The implicit progress made in Vienna was a start in teasing apart these three issues. Kennedy would hold his ground on West Berlin, but would also recognize Khrushchev's valid concerns about German rearmament. Kennedy made clear to Khrushchev in Vienna that he opposed "a buildup in West Germany that would constitute a threat to the Soviet Union."[46]

History has judged that Kennedy "lost" the Vienna Summit because he was browbeaten by Khrushchev, particularly when the two quarreled over ideology. Indeed, Kennedy himself described the Vienna experience as brutal, telling a *New York Times* reporter that it was the "worst thing in my life. He savaged me."[47] Yet Kennedy in fact had held firm and clear: the West would not buckle under threats on West Berlin. More important, and despite all appearances, Kennedy and Khrushchev were now on a path to resolve the larger German issues, a path they would pursue successfully in 1963.

But the next act on Berlin would be a public relations disaster for the Soviet Union. As tensions over Berlin mounted, more and more East Berliners voted with their feet by crossing the line to West Berlin and from there to the West generally. During the first six months of 1961, over 100,000 people left East Germany, with 50,000 fleeing in June and July alone.[48] The ongoing exodus of East Berliners was a huge economic loss to a floundering economy, but an even starker daily embarrassment in the frontline competition between socialism and capitalism. It was one of the main reasons why Khrushchev sought a long-term solution to Berlin.

As the flow of East Berliners accelerated, so did the geopolitical pressures. Finally, the East German government, with the support of its Soviet backers, moved to stanch the human flood. Berlin woke up on the morning of August 13 to a barbed wire divide, soon to be a concrete and heavily armed wall patrolled by around 7,000 soldiers and stretching 96 miles in Berlin and along the border between East and West Germany.[49]

Kennedy wisely did not challenge the Berlin Wall, except in a perfunctory manner, recognizing that any challenge or protest would prove empty. The West would certainly not go to war over Soviet actions on the *Soviet* side of the wall. Indeed, Kennedy immediately and correctly surmised that the wall might actually prove to be stabilizing in its perverse way, by removing the embarrassing and costly provocation of mass outmigration from East Berlin. As he suspected, the end of the flow of migrants from East Berlin rather quickly eased the Berlin crisis.

In fact, the overall Berlin situation seemed to Kennedy to be a dangerous snare that was certainly not worth the risks of global war. After the Vienna Summit, Kennedy had commented to his close aide Kenneth O'Donnell:

We're stuck in a ridiculous situation . . . God knows I'm not an isolationist, but it seems particularly stupid to risk killing a million Americans over an argument about access rights on an *Autobahn* . . . or because the Germans want Germany reunified. If I'm going to threaten Russia with a nuclear war, it will have to be for much bigger and more important reasons than that. Before I back Khrushchev against the wall and put him to a final test, the freedom of all Western Europe will have to be at stake.[50]

Kennedy would continue, successfully, to defend West Berlin, but he would also recognize the need to move to a sounder, long-term position with the Soviet Union vis-à-vis Germany. That insight was crucial to Kennedy's eventual success in 1963, and to the calming of the East-West confrontation thereafter. It was a basic strategic insight that Eisenhower never recognized or acted upon.

Chapter 2.

TO THE BRINK

The Cuban Missile Crisis

In the tense aftermath of the Vienna Summit, military budgets soared on both sides. A voluntary suspension of atmospheric nuclear testing that had lasted since 1958 fell apart in August 1961 when Khrushchev announced the Soviet Union would resume tests, breaking an explicit promise not to unilaterally resume tests that he had made to Kennedy two months earlier in Vienna. When Kennedy was told the news upon arising from a nap, his first reaction was "fucked again."[1] His closest advisers, McGeorge Bundy and Ted Sorensen, later said Kennedy felt more betrayed by the resumption of Soviet nuclear testing than by any other Soviet action during his presidency.[2] On October 30, the Soviet Union detonated a fifty-seven-megaton hydrogen bomb nicknamed the Tsar Bomba, which remains the most powerful nuclear weapon ever detonated.[3] After agonizing over the political ramifications, Kennedy reluctantly gave the order for the United States to resume testing as well. When Adlai Stevenson, former Democratic presidential nominee and Kennedy's ambassador to the United Nations, complained about the resumption of U.S. testing, Kennedy told him:

> What choice did we have? They had spit in our eye three times. We couldn't possibly sit back and do nothing . . . The Russians made two tests *after* our note calling for a ban on atmospheric testing . . . All this makes Khrushchev look pretty tough. He has had a succession of apparent victories—space, Cuba, the Wall. He wants to give out the feeling that he has us on the run . . . Anyway, the decision has been made. I'm not saying that it was the right decision. Who the hell knows?[4]

The two superpowers had entered an icy and dangerous phase of the Cold War, exactly the opposite of what both Kennedy and Khrushchev had hoped for at the start of 1961. In January 1962, when asked about the greatest disappointment of his first year in office, Kennedy answered: "Our failure to get an agreement on the cessation of nuclear testing, because . . . that might have been a very important step in easing the tension and preventing a proliferation of [nuclear] weapons."[5] The arms race, the Bay of Pigs, the confrontation in Vienna, and other events had eclipsed the early hopes.

Both countries pursued an aggressive agenda. Military budgets grew. Nuclear testing resumed. The Soviet ultimatum on Berlin was repeated, and the U.S. rejection of it held firm as well. In October 1961, Kennedy had his deputy secretary of defense, Roswell Gilpatric, give a blunt speech outlining America's nuclear superiority, significantly upping the ante. For the first time, the U.S. government publicly spelled out the vast nuclear advantage of the United States over the Soviet Union.[6] This was enormously revealing to the public, especially given Kennedy campaign rhetoric about the "missile gap."

Khrushchev was deeply undercut politically by the speech, as his domestic foes and competitors clearly recognized. Gilpatric's revelations called into question the Soviet capacity to make good on its threats. More fundamentally, the revelation of the U.S. nuclear advantage undermined Khrushchev's basic strategy of peaceful coexistence, by giving ammunition to Soviet hardliners who claimed that the Western alliance was out for a first nuclear strike against the Soviet Union. Khrushchev found himself increasingly cornered at home, in part through wounds self-inflicted by boasting and brashness, and in part through the arms race itself. He had overplayed his response to Kennedy's Bay of Pigs blunder, and found Kennedy to be a tougher foe than he had expected.

A furious Khrushchev decided on an even bolder stroke, a further escalatory move that in his estimation would address the nuclear imbalance, defend Soviet global interests, and present a stunning fait accompli to his U.S. counterpart. He would secretly place intermediate-range nuclear missiles in Cuba, and then reveal them to the world after the November 1962 U.S. congressional elections. He saw several advantages to this remarkable scheme. First, with missiles only ninety miles from the coast of Florida, the United States would feel the same pressure that the Soviet Union felt with U.S. nuclear missiles stationed nearby in Turkey and Italy. Second, the Soviet action would not require massive budgetary outlays, and therefore could be accomplished without a reversal of Khrushchev's desired strategy of diverting military spending to raising living standards domestically. Third, the Soviet Union would be capitalizing on the Bay of Pigs blunder by presenting the move as a defense of a fraternal socialist ally, in the context of a brazen failed invasion by the imperialist United States that suggested a high likelihood of future U.S. strikes against Cuba. Fourth, the Soviet Union would have a strong bargaining chip in any future negotiations over Berlin or other contested issues.[7]

The one thing that Khrushchev clearly did not want from this move was war. In explaining his proposal to his skeptical foreign minister, Andrei Gromyko, he emphasized that "[w]e don't need a nuclear war and we have no intention of fighting."[8] In winning the support of the Politburo for the move, Khrushchev repeatedly emphasized that "the Soviet Union should not and would not go so far as to risk a nuclear conflict."[9] The proposal was all about perceptions and threats, not about any desire to fight an actual nuclear war. But the reality was that a battle of perceptions could easily and accidentally slide into a shooting war, and a shooting war could become Armageddon.

Starting in August and escalating in October 1962, the Cuban Missile Crisis quickly unfolded.[10] Khrushchev's plan was to install the missiles by stealth and then unveil the new situation on a visit to the UN in November. In the meantime, Khrushchev would falsely maintain that any military support to Cuba would be conventional and defensive in nature. He naïvely believed that the nuclear missiles would be in place before the United States noticed, despite ongoing U.S. surveillance. At that point the fait accompli would be irreversible. If the United States could have nuclear missiles pointed at the Soviet Union from nearby bases in Turkey and southern Europe, the United States would be forced by world opinion, Khrushchev believed, to accept the Soviet missiles in Cuba.

A massive flotilla of Soviet ships carrying the missiles, other military equipment, and Soviet army units set sail for Cuba. The United States quickly noticed the increase of sea shipments, and Kennedy ordered the CIA to step up U-2 aerial surveillance and other intelligence gathering (including using Cuban informants). Several Republican senators, led by Kenneth Keating, berated the administration for allowing a Soviet weapons buildup in Cuba, and asserted that the Soviet Union planned to introduce nuclear missiles into the new military installations. Kennedy and most of his advisers rejected this claim as political theater, refusing to believe that Khrushchev would ever implement such a dangerous and destabilizing move. Kennedy publicly declared in September that the United States would never tolerate offensive Soviet weapons in Cuba, promising that "if Cuba should possess a capacity to carry out offensive actions against the United States . . . the United States would act."[11] He also denied that such a buildup was taking place, relying on repeated Soviet assurances that all weapons being introduced into Cuba were purely defensive in nature.

On October 15, further U-2 surveillance revealed the shocking truth: offensive weapons, both missiles and fighter planes, were in Cuba, with more on their way, in direct contradiction to repeated Soviet promises. Moreover, the installations were far along, though the surveillance could not say with assurance exactly how far. Perhaps some of the missiles were already installed. Others might be ready to launch in hours or days.

The next thirteen days were the most perilous in the planet's history. The drama was largely captured on White House tape recordings and is now, of course, the subject of a vast and detailed recounting. Kennedy put the odds of a nuclear war at "somewhere between one out of three and even."[12] He immediately created an Executive Committee (ExComm) of the National Security Council, where the president; his brother, Attorney General Robert Kennedy; and around a dozen senior officials huddled for agonizingly tense days, trying to plot a way out of the crisis.* At the very start of the ExComm process, Kennedy made the basic decision—one that was never second-guessed within the group—that the Soviet weapons must go. Either the weapons would be removed peacefully by the Soviets themselves, or they would become the cause of war.

---

* Key members of the ExComm included Vice President Lyndon B. Johnson, Secretary of State Dean Rusk, Secretary of Defense Robert McNamara, Secretary of the Treasury Douglas Dillon, Attorney General Robert Kennedy, National Security Adviser McGeorge Bundy, CIA director John McCone (who was one of the earliest officials to accurately warn of the threat of Soviet missiles in Cuba, but whose assessment was undercut by his subsequent departure for his honeymoon), Joint Chiefs of Staff chairman General Maxwell Taylor, Undersecretary of State George Ball, ambassador to the Soviet Union Llewellyn Thompson, Deputy Secretary of Defense Roswell Gilpatric, former secretary of state Dean Acheson, special assistants to the president Kenneth O'Donnell, David Powers, and Theodore Sorensen, and adviser Paul Nitze.

ExComm meeting during the Cuban Missile Crisis. Clockwise from President
Kennedy (center): Secretary of Defense Robert S. McNamara; Deputy Secretary
of Defense Roswell Gilpatric; chairman of the Joint Chiefs of Staff General Maxwell Taylor; Assistant Secretary of Defense Paul Nitze; Deputy Director of the
U.S. Information Agency Donald Wilson; special counsel Theodore Sorensen;
special assistant McGeorge Bundy; Secretary of the Treasury Douglas Dillon;
Attorney General Robert F. Kennedy; Vice President Lyndon B. Johnson (hidden); Ambassador Llewellyn Thompson; Arms Control and Disarmament
Agency director William C. Foster; CIA director John McCone (hidden); Undersecretary of State George Ball; Secretary of State Dean Rusk (October 29, 1962).

Several considerations went into this unanimous view. Most
important was Kennedy's belief that the United States would
suffer a decisive loss of international power and prestige if the
missiles remained in place. Kennedy had clearly and irrevocably
drawn a line just a month earlier: the United States would not tolerate Soviet offensive weaponry in Cuba. He could not afford to
lose face by abandoning that commitment. As he commented
during the deliberations, "Last month I should have said that we
don't care. But when we said we were not going to [allow missiles
in Cuba] and they go ahead and do it, and then we do nothing, I
would think that our risks increase."[13]

Second was Kennedy's domestic political flank. The Republican
right and some senators in his own party believed that Kennedy

was too soft in his foreign policy, a viewpoint that was heightened after Kennedy's reluctance to use U.S. military force in the Bay of Pigs invasion. If he failed to act now, his domestic base of support might collapse. During the crisis, he asked his brother Robert why they were risking war; Robert replied, "I just don't think there was any choice . . . if you hadn't you would have been impeached." Kennedy responded, "That's what I think, I would have been impeached."[14]

The ExComm held divergent views on the substantive effect of the missiles on the East-West *military* balance. Secretary of Defense Robert McNamara held that the missiles had zero net effect, given that the Soviets had intercontinental ballistic missiles (ICBMs) that could target the United States from Soviet territory anyway. The military brass felt otherwise, that Soviet missiles just off the U.S. coast would substantially enhance Soviet military power, especially since the Soviet strategic forces at that point depended overwhelmingly on bombers with a long and difficult flight path to the United States. All agreed, however, that the missiles must go.

At the start of the secret deliberations, the president himself and almost all of the advisers, particularly the Joint Chiefs of Staff, advocated a U.S. military air strike on the Soviet installations—an action that would have guaranteed a shooting war on both sides, particularly given that the military couldn't guarantee complete accuracy in destroying all the Cuban missiles. It might also have triggered a Soviet retaliation in Berlin, and perhaps in additional hotspots. Nor could one rule out the escalation from an anti-missile attack to an all-out nuclear Armageddon. Here then was the bind. The missiles needed to go, a military option seemed necessary, yet that option could trigger mutual destruction.

The ExComm process offered one absolutely decisive benefit: time. The ExComm was able to sit together in secret all-day meetings, debating options and strategies. This forced a thorough review of options, and it allowed some time for communication between Kennedy and Khrushchev, albeit through a laborious and confused process of letters, public pronouncements, telegrams, and messengers. It gave time for heated emotions—panic, fear, and desire to lash out at the adversary—to be kept in check so that reason could be invoked. "Slow" rational thinking was given time to dominate the "quick" emotional thinking.* And ultimately, it gave time for negotiations to ensue. Although most of the generals continued to push for a quick military strike, the president, his brother, Sorensen, and several other advisers eventually came to favor a much more gradual approach. This started with a naval quarantine of Soviet ships entering Cuban waters, announced publicly and sanctioned by the Organization of American States, combined with a U.S. demand that the missiles be removed, with any shooting held off as a last resort. These tactics proved critical. The quarantine gave Khrushchev time to consider his moves and the opportunity to pull back.

As with his demands over Berlin, Khrushchev had to stand down, this time in full view of the world. Kennedy was determined, however, to help Khrushchev save face. Most important, Kennedy agreed to Khrushchev's public demand to combine the Soviet withdrawal of the missiles with a U.S. pledge not to invade Cuba. Kennedy would secure the dismantling of the missile installations, while Khrushchev could at least claim to have saved Cuba from another attempted U.S. invasion.

Privately, Kennedy also substantially sweetened the deal by letting Khrushchev know that within five to six months of the dismantling of the Cuban missiles, the United States would withdraw its own Jupiter missiles from Turkey, a condition that Khrushchev had publicly called for in a radio address. Kennedy took this step on three conditions. First, the Turkish missile withdrawal could

---

* Psychologist Daniel Kahneman has described the distinctive mental pathways of "slow" rational thought in the prefrontal cortex and "fast" emotional thought involving other parts of the brain. Daniel Kahneman, *Thinking, Fast and Slow* (New York: Farrar, Straus and Giroux, 2011).

not be viewed as a quid pro quo, as Khrushchev's phrasing had implied. In fact, Kennedy mildly prevaricated by telling Khrushchev that the decision to remove the Jupiter missiles had already been made before the Cuban Missile Crisis. Second, the formal decision would be NATO's, to be confirmed later, rather than that of the United States alone. And third, it would have to be kept confidential, in order to avoid the appearance that the United States had struck a bargain at the expense of an ally. Any leak of the decision would force the United States to cancel the withdrawal of the Jupiter missiles. Kennedy entrusted his closest confidant, his brother and attorney general, Robert, to deliver this message in a secret meeting with Soviet ambassador Anatoly Dobrynin. Amazingly, the secrecy held on all sides for more than twenty years. Even most members of ExComm remained in the dark afterward, not realizing that the removal of U.S. missiles from Turkey had helped to resolve the crisis.*

A close reading of the timeline shows that Kennedy's concession was not actually necessary to end the crisis. Khrushchev had decided to withdraw the weapons before news of the Kennedy offer on the Jupiter missiles reached him. Nonetheless, Kennedy's concession surely paved the way for further progress in U.S.-Soviet relations. It revealed that Kennedy could see things from the Soviet perspective, and was prepared to act symmetrically. Khrushchev had indeed blinked, but Kennedy had demonstrated the constructive flexibility that would further the working relationship between the two leaders. In a private letter of October 28, the day the acute phase of the crisis concluded, Khrushchev expressed his great appreciation to Kennedy:

---

* Once this deal was made public, adviser McGeorge Bundy, one of the few who knew about it, noted, "There was no leak. As far as I know, none of the nine of us told anyone else what had happened. We denied in every forum that there was any deal." McGeorge Bundy, *Danger and Survival: Choices About the Bomb in the First Fifty Years* (New York: Random House, 1988).

Agreement on this matter, too, shall be a no [sic] small step advancing the cause of relaxation of international tensions and the tensions between our two powers. And that in turn can provide a good impetus to resolving other issues concerning both the security of Europe and the international situation as a whole.[15]

This is a notable statement by Khrushchev, underscoring the wisdom of Kennedy's action. And Khrushchev never revealed the secret of Kennedy's Jupiter trade, even while he was lambasted at home and abroad for a humiliating retreat under U.S. pressure. The Soviet military soon began dismantling and crating the weapons in Cuba for return to the Soviet Union. Kennedy strongly admonished his jubilant and relieved team to avoid any public gloating. The president's public standing soared, contributing to his party's successful congressional election results the next week. Yet opponents from the extremes also attacked the peaceful outcome. Castro denounced the agreement as an abandonment of Cuba and socialist resolve. The communist People's Republic of China mocked Khrushchev for appeasing the West, denigrating the U.S. nuclear threat as a "paper tiger" (to which Khrushchev reportedly answered that the tiger had "nuclear teeth").[16] The American right bemoaned the new commitment to not invade Cuba. Senator Barry Goldwater, who would become the 1964 Republican nominee for president, complained vociferously that the no-invasion pledge had "locked Castro and communism into Latin America and thrown away the key to their removal."[17]

Picking Up the Pieces

The world had never before peered into the abyss as it did in those days. And the two leaders who steered the world away from it,

Kennedy and Khrushchev, were now joined by this near-death experience, each feeling a responsibility that only the other could
understand. As Kennedy said, "The president bears the burden of
responsibility. The advisers may move on to new advice."[18] David
Ormsby-Gore, the British ambassador to the United States and a
close Kennedy friend for decades, recalled that after the crisis, "he
finally realized that the decision for a nuclear holocaust was his.
And he saw it in terms of children—his children and everybody
else's children. And then that's where his passion came in, that's
when his emotions came in."[19] Khrushchev also recalled the terror
of the missile crisis: "Any man who could stare at the reality of
nuclear war without sober thoughts was an irresponsible fool . . .
Of course I was scared. It would have been insane not to have
been scared. I was frightened about what could happen to my
country—or your country and all the other countries that would
be devastated by a nuclear war."[20]

Both leaders were changed and sobered by the events. Both
realized how the world was on a hair trigger, how misunderstanding could lead to utter disaster, and how fragile their positions had
been during the crisis. For those thirteen days, local commanders
on both sides could easily have sparked a global war by disobey-
ing or misunderstanding orders from above, or by acting on the
prerogatives that they were granted as a result of the heightened
military alert status. And despite all efforts by both sides to
avoid calamitous accidents, such calamities nevertheless nearly
occurred multiple times. Khrushchev was furious when a local
Soviet unit in Cuba shot down a U-2 spy plane at the most delicate
moment in the negotiations.

Kennedy's mistrust of the military grew even stronger after the
Cuban Missile Crisis, during which the generals had single-
mindedly advocated military strikes that almost certainly would
have led to nuclear war. Robert Kennedy concluded that "this experience pointed out for all of us the importance of civilian direction and control," and the president concurred.[21] The crisis further

strengthened Kennedy's confidence in his own foreign policy judgment, and his resolve to more forcefully steer the course of relations in the future.

Kennedy concluded from the crisis that a "world in which there are large quantities of nuclear weapons is an impossible world to handle."[22] He had come to the view that proposals to share nuclear weapons with NATO under a Multilateral Force needed to be shelved despite pressure from U.S. diplomats and West German Chancellor Adenauer. The nuclear arsenal of the Western alliance would be limited to the United States, the United Kingdom, and France, without Germany. And weighing on his mind was the fact that the People's Republic of China would soon have its own bomb, with many others to follow. Khrushchev had the same instinct. He wrote to Kennedy that they should use the aftermath of the crisis to end nuclear tests "once and for all."[23]

Kennedy and Khrushchev realized they shared the responsibility to avoid blowing up the world. Khrushchev emphasized that the missiles in Cuba did not in any way signal a desire for nuclear war:

> Only lunatics or suicides, who themselves want to per-ish and to destroy the whole world before they die, could do this [attack you]. We, however, want to live and do not at all want to destroy your country . . . Our view of the world consists in this, that ideological ques-tions, as well as economic problems, should be solved not by military means, they must be solved on the basis of peaceful competition.[24]

The correspondence between the two leaders in the months that followed shows how the prospects for an agreement had markedly brightened. Some historians argue that this was because Khrushchev had lost the battle with Kennedy and now had to give ground. I think rather it was because both leaders recognized their shared desire to avoid war, and their stark realization that many of

the generals and advisers on both sides had urged a military response that would have threatened all human survival. Khrushchev
praised Kennedy for holding back the more militaristic views:

> You evidently held to a restraining position with regard
> to those forces which suffered from militaristic itching.
> And we take a notice of that. I don't know, perhaps I am
> wrong, but in this letter I am making the conclusion
> on the basis that in your country the situation is such
> that the decisive word rests with the President and if he
> took an extreme stand there would be no one to restrain
> him and war would be unleashed. But as this did not
> happen and we found a reasonable compromise hav-
> ing made mutual concessions to each other and on this
> basis eliminated the crisis which could explode in the
> catastrophe of a thermonuclear war, then, evidently,
> your role was restraining. We so believe, and we note
> and appreciate it.[25]

Both sides, therefore, began to step back from the abyss, determined to expand the peaceful resolution of the crisis into longer-
lasting diplomatic results. This was a chance for the United States
and the Soviet Union to conclude at long last the negotiations on
arms control that had proven elusive for nearly a decade. It would
be Khrushchev and Kennedy's culminating work—Khrushchev in
his final full year of power, Kennedy in his last year of life.

Chapter 3.

PRELUDE TO PEACE

K E N N E D Y   H A D   C O M E to office inexperienced. He had a lot to learn about presidential leadership, and he needed time and experience to assess the quality of the advice he was receiving. He also needed time to build up his own credibility. He would have to face down a lot of domestic opposition to succeed in negotiations with the Soviet Union, given the previous fifteen years of confrontation, arms race, and anti-Soviet rhetoric. From the start, Kennedy composed his foreign policy team of establishment heavyweights—McGeorge Bundy, Robert McNamara, Douglas Dillon, Dean Rusk, Averell Harriman, Paul Nitze, and others—yet it took time to make them a team, and indeed his team.* He also suffered the powerful holdovers from the political right, even the

---

* Key members of his foreign policy team included McGeorge Bundy, national security adviser; Robert McNamara, secretary of defense; Dean Rusk, secretary of state; Averell Harriman, former ambassador to the Soviet Union and Kennedy's assistant secretary of state for Far Eastern affairs (November 1961–April 1963) and undersecretary of state for political affairs (April 1963–March 1965); Carl Kaysen, deputy special assistant for national security affairs; Roswell Gilpatric, deputy secretary of defense; and Paul Nitze, assistant secretary of defense for international security affairs.

Kennedy meets with State Department officials. Far side of table: U.S. ambassador to the Soviet Union Llewellyn Thompson; Vice President Lyndon B. Johnson; ambassador at large Averell Harriman; special assistant to the secretary of state Charles Bohlen. With backs to camera, left to right: ambassador-designate to Yugoslavia George Kennan; President Kennedy; Secretary of State Dean Rusk (February 11, 1961).

far right, including CIA director Allen Dulles, CIA deputy director Richard Bissell (both of whom Kennedy forced to resign after the Bay of Pigs debacle), and J. Edgar Hoover at the FBI. And Kennedy himself lacked experience in executive organization, foreign policy leadership, and dealing with both the Soviet Union and his own noisy and opportunistic allies across Europe and Asia.

His first two years in office were marked by an unending series of crises: the Bay of Pigs, the Vienna Summit, the Berlin Wall, skirmishes in Laos and Vietnam, and the Cuban Missile Crisis. Kennedy learned quickly and grew markedly from these experiences. The Cuban Missile Crisis was the catharsis and turning point. From then to the end of his life a year later, Kennedy led. He became a master of events, not their pawn. He envisioned a pathway to peace, and achieved it. He was a changed man, and he changed the world.

Kennedy's Evolving Strategy of Peacemaking

In his final and commanding year, Kennedy implemented a strategy of peacemaking, one deeply grounded in both concept and experience. He was both idealist and realist, visionary and arm-twisting politician. The two approaches, one soaring and one with feet firmly on the ground, were necessary for success. He had mastered the double-barreled strategy he much admired in Churchill.

Kennedy had come to office with four basic precepts, to which he added a fifth and sixth. First, he had long recognized that the arms race under way in the 1950s and early 1960s was a prisoner's dilemma. Both sides had amassed nuclear weapons to the point of massive overkill, and the arms spiral was largely self-fulfilling. As Kennedy declared in the inaugural address:

> [N]either can two great and powerful groups of nations take comfort from our present course—both sides over-burdened by the cost of modern weapons, both rightly alarmed by the steady spread of the deadly atom, yet both racing to alter that uncertain balance of terror that stays the hand of mankind's final war.

The implication of the prisoner's dilemma is that there are large *mutual* gains from cooperation. Peace is worth pursuing for both sides. The Cold War was not a struggle in which the gains of one side equaled the losses of the other. The belief in mutual gains from peace is fundamentally different from the conception of the Cold War as a zero-sum struggle, a titanic fight to the death between competing ideologies, of God-fearing freedom versus athe-istic tyranny (in the view of U.S. hardliners), or of rapacious global capitalism versus historically ascendant communism (as it was seen by the Soviet hardliners).

The mutual gains, Kennedy felt, could be immense. Again, from the inaugural address:

> Let both sides seek to invoke the wonders of science instead of its terrors. Together let us explore the stars, conquer the deserts, eradicate disease, tap the ocean depths, and encourage the arts and commerce.
>
> And if a beachhead of cooperation may push back the jungle of suspicion, let both sides join in creating a new endeavor, not a new balance of power, but a new world of law, where the strong are just and the weak secure and the peace preserved.

Kennedy's second precept was that the arms race was not only costly, but also inherently unstable. The idea propounded by many nuclear strategists of a "stable balance of terror" was naïve. The rapid buildup of arms gives rise to a rapid buildup of risks as well, of accidents and unintended consequences, as the mishaps from the Bay of Pigs to the Cuban Missile Crisis amply demonstrated. When he became president, Kennedy frequently recalled the calamity of World War I, a war that resulted from a series of interlocking misjudgments and false perceptions among military and political leaders. By the end of the Cuban Missile Crisis he was horrified that he had nearly committed the same missteps.

In the vision of war by mishap, Kennedy had been deeply influenced, as we have seen, by Churchill's description of World War I in *The World Crisis,* a book that he had read as a fifteen-year-old and that weighed on his mind ever after. More recently, Kennedy had been moved by historian Barbara Tuchman's *The Guns of August,* a neo-Churchillian account that described the false premises, errors, miscalculations, and flawed military doctrines that led to the war.[1] Kennedy was so much influenced by it that he gave a copy to the secretary of the army, Elvis Stahr Jr., telling him, "I want you to read this. And I want every officer in the Army to read it." Stahr had the book placed in every officers' day room around the world, with a note saying that it came from the president.[2]

Kennedy had also been affected by the wisdom of British war theorist B. H. Liddell Hart, whose book *Deterrent or Defense* Kennedy had reviewed favorably for the *Saturday Review* during the 1960 campaign. Liddell Hart wrote:

> The study of war has taught me that almost every war was avoidable, and that the outbreak was most often produced by peace-desiring statesmen losing their heads, or their patience, and putting their opponent in a position where he could not draw back without serious loss of "face" . . .
>
> The best safeguard of all is for all of us to keep cool . . . War is not a way *out* from danger and strain. It's a way *down* into a pit—of unknown depth.
>
> On the other hand, tension so intense as it has been during the last decade [the 1950s] is almost bound to relax eventually if war is postponed long enough. This has happened often before in history, for situations change. They never remain static. But it is always dangerous to be too dynamic, and impatient, in trying to force the pace. A war-charged situation can only change in two ways. It is bound to become better, eventually, if war is avoided without surrender. Such logic has been confirmed by experience.[3]

Kennedy had adopted Liddell Hart's precept of patience as the guiding principle for managing the Cuban Missile Crisis. Find time. Delay. Allow the other side to do the same, and find a way for the aggressor to retreat with some dignity. Don't shoot first, to "get it over with." The Liddell Hart approach had worked. Kennedy was determined to keep this lesson in mind as he pursued a longer-lasting peace.

Kennedy was also becoming increasingly aware of technological breakdowns that could trigger a nuclear war that nobody wanted. The command-and-control systems governing complex military systems were highly vulnerable to breakdown. At the height of the Cuban Missile Crisis his orders to suspend U-2 flights inadvertently went unheeded by at least one pilot. He learned in horror how that pilot had then accidentally gone off course into Russian airspace. In addition, numerous phenomena caused false alerts of Soviet nuclear attacks, which might have triggered a disastrous response. These included the aurora borealis (the northern lights), space debris, a full moon, computer errors, and a Norwegian weather research rocket.[4]

Kennedy's third precept was that peace is a process, a series of step-by-step confidence-building measures. He recognized that moves by one side lead to moves by the other. A situation of high distrust necessitated a series of confidence-building steps. He and Khrushchev had seen the hard way that distrust on each side could quickly spiral out of control—even out of the leaders' control. Kennedy would repeatedly emphasize that success would occur one step at a time, and it was the responsibility of leaders, here and now, to take that first step.

Here is how he put this issue in April 1963 in his letter to Khrushchev, co-signed with British prime minister Harold Macmillan, proposing three-party talks on a test ban treaty:

> We know that it is argued [by Khrushchev himself] that a nuclear tests agreement, although valuable and welcome especially in respect of atmospheric tests, will not by itself make a decisive contribution to the peace and security of the world. There are, of course, other questions between us which are also of great importance; but the question of nuclear tests does seem to be one on which agreement might now be reached. The mere fact of an agreement on one question will inevi-

tably help to create confidence and so facilitate other
settlements.[5]

Kennedy's fourth precept was that peace must be pursued in
a manner that defends the fundamental interests of each side. In
the U.S. case, this meant resisting any rollback of democracy or,
say, the loss of West Berlin. Peace would be achieved through
cooperation, not through appeasement. The West would hold
its ground against communism. Negotiation through strength,
à la Churchill, and security through containment, as famously
outlined by diplomat George Kennan in 1947, remained Kennedy's
cornerstones. Arms control and the reduction of tensions would
be pursued while defending the core interests of the United States.

But how to cooperate, not appease? Striking this balance was
the critical step of Kennedy's maturation. At the start of his presidency, Kennedy had been too ready to accede to the generals, too
fearful of rebuke from the right, to accommodate legitimate Soviet
concerns and interests. Yet he learned to listen more clearly to
Khrushchev, and to see both sides of the security issues. He found
a point of equilibrium. Later he would describe this point as a spot
to place a lever that could move the world toward peace.

One of Kennedy's great strengths in finding this balance was his
recognition that the enmities of nations should not be viewed as
permanent. Cooperation toward shared strategic interests could
overcome deep historical and ideological differences even in the
most unlikely cases. The United States and the Soviet Union could
find points of agreement to unwind the Cold War. Kennedy would
repeat this theme many times in 1963, when he urged Americans
to imagine a world beyond the Cold War.

These four precepts were already reflected in Kennedy's inaugural address. They were deepened by experience and remained at
the core of Kennedy's approach to peacemaking. But Kennedy
added two crucial new precepts as the result of his first two years
as president.

Kennedy's fifth precept was that it is crucially important to listen carefully to the other side, given the inherent difficulties of accurate communication. His long correspondence with Khrushchev was critical in solving the Cuban Missile Crisis. It helped each side discern what was really important in their mutual dealings. Noise, propaganda, and public confusion were inevitably part of U.S.-Soviet relations. Kennedy and Khrushchev both recognized the value of their private communications, outside the glare and distortions of the media.

Kennedy deepened his communication with Khrushchev by using an informal go-between, *Saturday Review* editor and peace activist Norman Cousins. Cousins visited Khrushchev twice, in December 1962 and April 1963, carrying messages between Kennedy and Khrushchev, as well as serving as an informal emissary of Pope John XXIII to both leaders. Meeting Cousins in the White House in spring 1963, Kennedy described the great practical difficulties of clear communication, referencing the growing rift with intransigent ally France:

> You know, the more I learn about this business, the more I learn how difficult it is to communicate on the really important matters. Look at General de Gaulle. He's one of our allies. If we can't communicate with him and get him to understand things, we shouldn't be surprised at our difficulty with Khrushchev.[6]

What had Kennedy learned from his long communication with Khrushchev? He had learned, first, that Khrushchev and he faced the same problem in pursuing peace: the hardliners on their own teams. As he said to Cousins:

> One of the ironic things about this entire situation is that Mr. Khrushchev and I occupy approximately the same political positions inside our governments. He

would like to prevent a nuclear war but is under severe pressure from his hard-line crowd, which interprets every move in that direction as appeasement. I've got similar problems. Meanwhile, the lack of progress in reaching agreements between our two countries gives strength to the hard-line boys in both, with the result that the hard-liners in the Soviet Union and the United States feed on one another, each using the actions of the other to justify its own position.[7]

Second, he had come to understand and to appreciate the nature of Soviet concerns. Khrushchev's actions and threats on Berlin were a symptom of a deeper anxiety: the resurgence of German power after the devastating experiences of German aggression in the world wars. They were not a mere bluff, and still less a crude Soviet attempt at global conquest. They were, in a sense, cries of fear. This was not easy to recognize, since fear was manifested as threat and bluster.

Khrushchev raised the issue in his meetings with Cousins. Yes, Khrushchev acknowledged, the Soviet Union "could crush Germany in a few minutes. But what we fear is the ability of an armed Germany to commit the United States by its own actions. We fear the ability of Germany to start a world atomic war."[8]

Kennedy not only came to understand more clearly the nature of Soviet concerns, he acted upon them. He, too, was wary of German fingers on the nuclear trigger. Since Adenauer was pressing hard for this, Kennedy would have to confront the West German chancellor, his own ally, on this point. He did more than that. By making clear his displeasure with Adenauer's aggressive push for nuclear weapons, he helped other leading German politicians to ease the eighty-seven-year-old Adenauer out of power in October 1963, to be replaced by a far less truculent successor, Ludwig Erhard.

Kennedy's sixth precept followed from all that he had learned in navigating the many crises of his first two years, and this became the keystone to all the rest. Only strong and vigorous presidential leadership could deliver peace. That leadership was required not only in dealing with the Soviet Union, but also and perhaps especially in terms of the U.S. public, military and political elites, and European allies. Kennedy was not negotiating only with Khrushchev. He was constantly negotiating, maneuvering, and forging alliances at home and in Europe as well. Only active presidential leadership would overcome the doubts, fears, and provocations of the military, the hardliners, and the public. The Soviet leader faced the same constraints, perhaps even tougher ones. Both Kennedy and Khrushchev gave ground to each other to enable his counterpart to face down his own domestic skeptics and critics.

Through hard experience, Kennedy came to appreciate that only the president could set a vision of peace, and that only the president could overcome the deeply entrenched false assumptions held by the military, the foreign policy establishment, and the public, after years of anti-Soviet rhetoric and strategy. As the scholar James Richter noted, "Domestic politics of the great powers will also act as a brake on change . . . once established, [legitimating] myths become embedded in countries' domestic politics and are difficult to dislodge."[9] A consensus to cooperate with the Soviet Union would never coalesce on its own in the early 1960s. It would need to be forged by Kennedy himself.

Kennedy entered 1963 determined to lead the way to peace despite all the skepticism and barriers. He was determined to use the significant political capital that he had garnered in the Cuban Missile Crisis to that end. He had come to believe that his relationship with Khrushchev would help make an agreement possible. He also knew something crucial that the public did not. The Cuban Missile Crisis had been solved by negotiation, not by hardline bluster or militarism. It had involved an informal handshake and trust at the top between two adversaries.

The Push for Peace

Both Kennedy and Khrushchev personally seized the opportunity opened by the crisis. In late 1962, they began an intensive exchange of letters on reviving the on-again, off-again negotiations on a nuclear test ban, an agreement made all the more urgent given the evident progress of China in acquiring nuclear weapons. On December 19, Khrushchev wrote Kennedy that the time had arrived to "put an end once and for all" to nuclear weapons testing.[10]

Old unsettled issues—about both Germany and arms control—would threaten an agreement once again, causing both sides to question the goodwill, resolve, and ability of the other side to keep hardliners in check. As the past failures of U.S.-Soviet negotiations had vividly shown, there could never be unity on either side about the way forward. Only leadership at the top would suffice. The next nine months would be a drama of the two protagonists, Kennedy and Khrushchev. Kennedy's theory of peace and Khrushchev's quest for peaceful coexistence would face their ultimate test.

Chapter 4.

THE RHETORIC OF PEACE

KENNEDY AIMED TO conjure peace through the spoken word. This is how he understood great leadership. This is how he would turn his own personal courage and skills to the service of humanity. As a senator, Kennedy had risen to national prominence with his Pulitzer Prize–winning book of 1957, *Profiles in Courage,* which highlighted acts of political courage by eight senators.[1] The book was, at the core, a compendium of bold oratory throughout U.S. Senate history, speeches by politicians of such piercing eloquence and truth that they outlasted their own age and inspired later generations. During the summer of 1963, Kennedy made a series of speeches on peace that would likewise move his contemporaries—including his adversaries—and later generations.

In his efforts to rally public opinion to his side, Kennedy was acutely aware of one of the greatest foreign policy setbacks of the previous half century: President Woodrow Wilson's failure in 1919 to sway public and senatorial opinion to the cause of the League of Nations. Kennedy thus sought to stir the public not with false promises but with hard realism, not with balm but with responsible talk about the great stakes of making peace. He would follow Churchill's dictum of straight talk even if it was painful and politically dangerous. Kennedy would appeal to the people, so that peace would be a triumph of democracy itself.

The Power of Oratory

Kennedy was of course no stranger to the power of oratory. He had studied it since his youth, relished it, championed it, and aspired to greatness in it. His lifelong role model, in this arena as in so many others, was Churchill, whose incomparable rhetoric had helped save a civilization. If it was true at the Battle of Britain that "never in the field of human conflict was so much owed by so many to so few," then it was also probably true that never in the course of human conflict had so few words inspired so many to act so valiantly.[2] Kennedy had been present for the speeches in the British House of Commons when war was declared in 1939, and had been deeply affected by them, particularly by Churchill's words. Kennedy noted in April 1963 that Churchill "mobilized the English Language and sent it into battle."[3] Churchill's oratory influenced Kennedy in more ways than one. On the campaign trail in 1960, Kennedy hired a drama coach to improve his speaking, and biographer Richard Reeves related his practice sessions: "Home alone in Washington, he would put on a silk bathrobe, pour himself a brandy, light up a cigar, and speak along with records of Winston Churchill's greatest speeches."[4] Kennedy himself would soon mobilize language in the battle for peace.

Kennedy had an important weapon, his own verbal Excalibur, a linguistic sword of unparalleled strength and balance, able to cut to the core of an issue with remarkable insight and eloquence. This was Ted Sorensen, trusted counselor, speechwriter, adviser, and according to most, Kennedy's intellectual alter ego. As Kennedy's

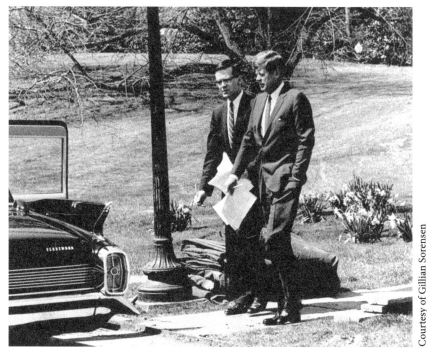

Courtesy of Gillian Sorensen

Kennedy with Ted Sorensen.

top national security adviser, McGeorge Bundy, noted, "There just is no exaggerating [Sorensen's] value and his closeness to the President . . . he had a deep sense of the President's own values and purpose."[5] Sorensen was Kennedy's chief speechwriter throughout his eight Senate years. He helped Kennedy assemble and write *Profiles in Courage*. He was Kennedy's wordsmith throughout the 1960 campaign and in the White House. And he was on hand as adviser, scribe, and speechwriter for every major challenge that Kennedy confronted, drafting the key documents that ended the Cuban Missile Crisis and those that would lead the final public campaign for peace.

Sorensen embodied virtues and values ideally suited for the cause of peace. He was a midwesterner, a son of Nebraska, with its great tradition of anti-war sentiment. At Sorensen's urging, Kennedy featured Senator George Norris's principled and brave opposition to the U.S. entry into World War I in *Profiles in Courage*. Sorensen was the son of a Jewish mother and Unitarian father—

a "Jewnitarian," as Sorensen joked in his memoirs. Both faiths shared the belief, in Sorensen's words, that "human beings represent the one true God here on Earth, and that good works by man are sanctifying God's name."[6] We can hear echoes of that faith in the closing words of Kennedy's inaugural address: "With a good conscience our only sure reward, with history the final judge of our deeds, let us go forth to lead the land we love, asking His blessing and His help, but knowing that here on earth God's work must truly be our own."

Kennedy and Sorensen teamed up on the great Peace Speech and the others that followed later that summer. While we don't know all of the sources that they drew upon, they would likely have reviewed several important speeches on peace given since the end of World War II, including Churchill's "Sinews of Peace," delivered at Westminster College in Fulton, Missouri, in March 1946, and three by Eisenhower: "Chance for Peace," delivered in April 1953, the month following Stalin's death; the "Atoms for Peace" speech at the UN in December 1953; and his farewell address, given three days before the end of his presidency in January 1961. As for Kennedy's own previous speeches, two had set the stage for the peace initiative: the inaugural address, of course, and also his first speech to the UN General Assembly in September 1961.

One final and enormous rhetorical influence arrived just before Kennedy's speech. It came not as a speech, but as a written message from the beloved Pope John XXIII, in his dying days. The pope had sent a message of peace during the darkest days of the Cuban Missile Crisis, and Khrushchev had expressed his personal gratitude to the pope for that message. That in turn encouraged the pope to devote his final days to the encyclical *Pacem in Terris* (Peace on Earth), which moved the global public, particularly the pope's message that "any disputes which may arise between nations must be resolved by negotiation and agreement, and not by recourse to arms."[7] The words of that inspiring encyclical, issued on April 11, 1963, just two months before Kennedy's Peace Speech, likely informed Kennedy's own speech.

"Sinews of Peace" (March 5, 1946)

Churchill's words were the primary model for Kennedy: vigorous, down-to-earth, a clarion call. Churchill's words informed, explained, called for action, and predicted success. He was a moralist, realist, activist, and visionary. After a Churchill speech, there was no choice but to sally forth, to do the deeds that would win the affection and honor of later generations. Churchill's rhetorical power came above all from the sense of fierce realism that he conveyed. He would not hold back vital information from compatriots and allies. His listeners knew this, and felt empowered by it.

In Fulton, Churchill had a few specific purposes.[8] He bid his listeners that day, including President Harry Truman, to grasp the reality of the newly divided Europe. Using the phrase that came immediately to define the Cold War, he sounded the alarm: "From Stettin in the Baltic to Trieste in the Adriatic an iron curtain has descended across the Continent." "Iron curtain" had an earlier provenance, having appeared in diplomatic, literary, and political references during the preceding half century; Churchill himself used it on several occasions in 1945. But it was the "Sinews of Peace" speech that defined the new era in those terms for the entire world.

Churchill's speech emphasized peace as well as the Cold War. It made clear that "we understand the Russian need to be secure on her western frontiers by the removal of all possibility of German aggression." Churchill underscored, "I do not believe that Soviet Russia desires war. What they desire is the fruits of war and the indefinite expansion of their powers and doctrines." He urged the United States and the United Kingdom to meet that expansionist tendency together firmly but peacefully.

Churchill's urgent message on that day was more about will-power than military might. He called for a special relationship between the United States and the British Commonwealth and Empire. He spoke of that relationship in the terms of everyday life, not of high diplomacy. "If two of the workmen [building the Temple of Peace] know each other particularly well and are old friends, if their families are intermingled . . . why cannot they work together at the common task as friends and partners?"

He called for building the new United Nations, so that "its work is fruitful, that it is a reality and not a sham, that it is a force for action, and not merely a frothing of words, that it is a true temple of peace in which the shields of many nations can some day be hung up, and not merely a cockpit in a Tower of Babel." And under the general authority of the UN he called for a new relationship with Russia, not at a vague point in the future, but then and now, in 1946. We should reach, said Churchill, "a good understanding on all points with Russia under the general authority of the United Nations Organization and by the maintenance of that good understanding through many peaceful years, by the whole strength of the English-speaking world and its connections."

Churchill ended his remarks to his American listeners with a challenge, a promise, and a vision. Our efforts could redeem the future, create the peace we sought:

> If we adhere faithfully to the Charter of the United Nations and walk forward in sedate and sober strength seeking no man's land or treasure, seeking to lay no arbitrary control upon the thoughts of men; if all British moral and material forces and convictions are joined with your own in fraternal association, the highroads of the future will be clear, not only for our time, but for a century to come.

Churchill's promise—peace "not only for our time, but for a century to come"—not only referenced Chamberlain's debacle eight years earlier in declaring the attainment of "peace in our time" after Munich, but looked forward confidently to a new era of peace. It was a turn of phrase that John Kennedy would echo seventeen years later in his Peace Speech, when he declared that we seek "not merely peace in our time but peace in all time."

Churchill would rue the fact that "Sinews of Peace" was later remembered mainly as a call to the Cold War, as its principal effect in the United States was to warn the public about the new "iron curtain." Yet the speech was vivid and unmistakable in its call for peace through strength:

> The safety of the world, ladies and gentlemen, requires a new unity in Europe, from which no nation should be permanently outcast. It is from the quarrels of the strong parent races in Europe that the world wars we have witnessed, or which occurred in former times, have sprung . . . Surely we should work with conscious purpose for a grand pacification of Europe, within the structure of the United Nations and in accordance with our Charter. That I feel opens a course of policy of very great importance.

Even if the world would remember the speech mostly for its grim warning, Kennedy certainly would remember it also for the purposes that Churchill intended: as a call for peace through strength, resolve, and negotiation.

"Chance for Peace" (April 16, 1953)

Churchill's speech was followed not by peace but by the Cold War, which began in earnest and rapidly spiraled into a series of crises: instability in Greece and Turkey and the Truman Doctrine in 1947; the Marshall Plan and its rejection by the Soviet Union the same year; the Berlin crisis and airlift in 1948–1949; the Soviet atomic bomb in 1949; the Chinese Communist victory over the Nationalists in 1949; the Korean War from 1950 to 1953; and many more. The first major pause of that spiral came with Stalin's death on March 5, 1953. Stalin's ruthlessness, brutality, and paranoia had fueled much of the Cold War in its early years, in addition to terrorizing his own colleagues and nation. His death opened the possibility of a thaw, if not a new start entirely. Eisenhower took a step toward that thaw with an important speech one month after Stalin's death, billed as the "Chance for Peace."*[9]

The speech was unusual for Eisenhower: a relatively bold attempt to defuse the Cold War dynamics. Eisenhower made a bid for a new relationship with the Soviet Union, but it was circumscribed, reflecting the doubts and cautions of Eisenhower's own hardline advisers, especially Secretary of State John Foster Dulles.

The opening line is dramatic: "In this spring of 1953 the free world weighs one question above all others: the chance for a just peace for all peoples." He recalled "the more hopeful spring of 1945, bright with the promise of victory and freedom," a moment of hope that he himself had helped to make possible as the victorious supreme Allied commander of World War II.

Eisenhower then went on to outline the "two distinct roads" charted by the United States and the Soviet Union. The United States, by Eisenhower's light, charted a path that was faithful to the

---

* Eisenhower had wanted to make such a speech for some time. Though the rhetoric in the final version was slightly toned down, he told his speechwriter Emmet Hughes that the essence of what he wanted to say was "The jet plane that roars over your head costs three quarters of a million dollars. That is more money than a man earning ten thousand dollars every year is going to make in his lifetime . . . We are in an armaments race: everyone is wearing himself out to build up his defenses. Where is it going to lead us? At worst, to atomic warfare, and we can state pretty damn plainly what that means. But at the least, it means that every people, every nation on earth is being deprived of the fruits of their own toil . . . Now, here's the other choice before us, the other road to take— the road of disarmament. What does that *mean*? It means for everybody in the world: butter, bread, clothes, hospitals, schools—good and necessary things for a decent living" Evan Thomas, *Ike's Bluff: President Eisenhower's Secret Battle to Save the World* (New York: Little, Brown, 2012), 60.

spirit of the UN: "to prohibit strife, to relieve tensions, to banish fears." The Soviet Union, by contrast, pursued a goal of "power superiority at all cost," in turn compelling the free nations to "remain armed, strong, and ready for the risk of war." The result was a way of life "forged by eight years of fear and force." On the current course, said Eisenhower, the best outcome is more fear and tension; the worst, atomic war.

Eisenhower bemoaned the huge costs for both sides:

> Every gun that is made, every warship launched, every rocket fired signifies, in the final sense, a theft from those who hunger and are not fed, those who are cold and are not clothed. This world in arms is not spending money alone.
>
> It is spending the sweat of its laborers, the genius of its scientists, the hopes of its children.
>
> The cost of one modern heavy bomber is this: a modern brick school in more than 30 cities.
>
> It is two electric power plants, each serving a town of 60,000 population.
>
> It is two fine, fully equipped hospitals. It is some 50 miles of concrete highway . . .
>
> This is not a way of life at all, in any true sense. Under the cloud of threatening war, it is humanity hanging from a cross of iron.

Eisenhower posed the key question that "stirs the hearts of all sane men: is there no other way the world may live?" And he asked the question of the new Soviet leadership. Will it "awaken, with the rest of the world, to the point of peril reached and help turn the tide of history"? He acknowledged that "[w]e do not yet know." Eisenhower said that only deeds would tell, giving the list: an Austrian peace treaty; the release of prisoners held since World War II; an honorable armistice in Korea; an end to the direct and indirect attacks on the security of Indochina and Malaya (the colonial possessions of Southeast Asia); and the fostering of a broader European community with a free and united Germany and the full independence of the East European nations.

All of this would make possible vast rewards, including an agreement on arms reduction and the enormous savings that would result. The possibilities of such an agreement would present the world "with the greatest task, and the greatest opportunity, of all. It is this: the dedication of the energies, the resources, and the imaginations of all peaceful nations to a new kind of war. This would be a declared total war, not upon any human enemy but upon the brute forces of poverty and need."

Eisenhower made a pledge:

This government is ready to ask its people to join with all nations in devoting a substantial percentage of the savings achieved by disarmament to a fund for world aid and reconstruction. The purposes of this great work would be to help other peoples to develop the undeveloped areas of the world, to stimulate profitable and fair world trade, to assist all peoples to know the blessings of productive freedom.

The offer was bold, but also limited in a basic way that Kennedy's offer ten years later would not be. As Eisenhower himself put it, he was offering the Soviet Union a test of good faith through a list of measures the Soviet Union must enact. "The test is clear," he said of Soviet actions. Kennedy, by contrast, would bid Americans to reexamine their own attitudes to peace, the better to meet the Soviet Union on common ground.

Eisenhower regarded this speech as one of his most important. In his 1961 memoirs, Eisenhower's chief of staff, Sherman Adams, declared it to be the greatest speech of Eisenhower's career. Unusually, the Soviet Union allowed its publication in full in the newspaper *Pravda,* showing the high regard that it gave to Eisenhower's peace gesture.[10]

The optimism lasted through the end of the year, when Eisenhower spoke to the UN General Assembly, offering "atoms for peace."*[11] There, Eisenhower spoke of the horrors of nuclear war and the horrendous consequences of the nuclear arms race. Both sides can inflict hideous damage on the other. And the nuclear stalemate was little reprieve, said Eisenhower: "To pause there [with devastating weapons pointed at each side] would be to confirm the hopeless finality of a belief that two atomic colossi are doomed malevolently to eye each other indefinitely across a trembling world." Kennedy would echo these words a decade later when he also challenged the view that we are "doomed," condemned to war.

Eisenhower called for patience in moving the world "out of the dark chamber of horrors into the light." He called for a step-by-step process of agreements on disarmament. And specifically, he suggested that both sides allocate a fraction of their fissionable material "to serve the peaceful purposes of mankind," for agriculture, medicine, and electrical power.

Yet the hopes of 1953 proved evanescent. The Soviet Union was torn by an internal power struggle, pitting Khrushchev against his competitors for power, Lavrentiy Beria, Georgy Malenkov, and Vyacheslav Molotov. Khrushchev won the backing of Soviet generals in part by urging a hardline response to the United States. And equally important, Eisenhower's own side pushed back, especially Secretary of State Dulles. Both sides missed the most important opportunity since the end of World War II to halt the spiraling instability and escalation.

Prime Minister Churchill met with Eisenhower in Bermuda in

---

* The drafting of the "Atoms for Peace" speech was a huge collaborative effort, spanning almost nine months from the time of Stalin's death to the day of Eisenhower's speech at the UN on December 8. The project was known as "Operation Wheaties" because the participants held breakfast meetings to discuss it. Meena Bose, *Shaping and Signaling Presidential Policy: The National Security Decision Making of Eisenhower and Kennedy* (College Station: Texas A&M University Press, 1998), 83.

December 1953 to urge a new peace initiative. Convinced that Stalin's death offered a rare opportunity to wind down the Cold War, Churchill told the House of Commons on May 11, 1953, that "[i]t would be a mistake to assume that nothing can be settled with Soviet Russia unless or until everything is settled."[12] Yet Eisenhower was more skeptical. He told Churchill rather brutally that Russia "was a woman of the streets and whether her dress was new, or just the old one patched, it was certainly the same whore underneath," demonstrating his deep hesitation about negotiating with the Soviets despite his desire for peace.[13] Churchill was dismayed to watch the United States squander this opportunity under the thrall of simplistic Cold War ideology and the dogma of "massive retaliation."

Nevertheless, Eisenhower longed to be a peacemaker. Given the severe limitations in mutual understanding between the United States and the Soviet Union, however, his moment never arrived. Stubborn disagreements between the United States and the Soviet Union over whether onsite inspections were required to distinguish underground nuclear explosions from earthquakes (as the United States held) prevented the completion of a much-discussed test ban treaty. And as so often happened in the Cold War, even when events were moving slowly in the right direction, they were knocked off track by miscalculation: Eisenhower's last push for peace in 1960 ended in the flames of Gary Powers's downed U-2 spy plane.

Eisenhower's Farewell Address
(January 17, 1961)

Eisenhower delivered his most famous speech just three days before he left the White House to John Kennedy.[14] Only two presidential farewell addresses are widely remembered today. George Washington used his to warn Americans about "entangling alliances overseas." Eisenhower used his to warn Americans about entangling alliances right at home, specifically those among the military, industry, and the government. Eisenhower's warning about the risk of the "military-industrial complex" has reverberated through a half century.

How poignant and powerful for America's greatest twentieth-century general to caution America about the threat the military posed to American democracy. No one besides Eisenhower could have had the stature and credibility to deliver this message:

> In the councils of government, we must guard against the acquisition of unwarranted influence, whether sought or unsought, by the military-industrial complex. The potential for the disastrous rise of misplaced power exists and will persist.
>
> We must never allow the weight of this combination to endanger our liberties or democratic processes. We should take nothing for granted. Only an alert and knowledgeable citizenry can compel the proper meshing of the huge industrial and military machinery of defense with our peaceful methods and goals, so that security and liberty may prosper together.

These words were prescient, and also bittersweet. Eisenhower had wanted to make peace with his Soviet counterparts, but he never succeeded. Events, experts, the CIA, and cabinet members always put obstacles in his path. He himself was perhaps too skeptical and detached. Eisenhower warned of the military-industrial complex, but he never really held it in check, or found the voice and presidential direction to achieve agreement with the Soviet Union.

Kennedy's Inaugural Address
(January 20, 1961)

Kennedy would draw on these precedents of public persuasion, and especially on Churchill's concepts and phrases, in his own quest for peace in 1963. He would talk about the mutual gains from disarmament; the need to stop the upward ratcheting of risks; the danger of accidents; and the interests of both sides in peace. Yet Kennedy also made important innovations in rhetoric and strategy, innovations that reflected his hard-won insights and personal courage.

Kennedy spoke of the quest for peace on countless occasions, starting with the powerful words of his inaugural address on January 20, 1961.[15] In his first moments as president, he stated clearly the unique challenge and risk of the time. "[M]an holds in his mortal hands," said Kennedy, "the power to abolish all forms of human poverty, and all forms of human life." This echoed the paradox previously expressed forcefully by Churchill, that "[w]e, and all nations, stand, at this hour in human history, before the portals of supreme catastrophe and of measureless reward. My faith is that in God's mercy we shall choose aright."[16]

Kennedy called on both sides to "begin anew the quest for peace, before the dark powers of destruction unleashed by science engulf all humanity in planned or accidental self-destruction." He then spelled out his strategy in an uncanny way. Echoing Eisenhower's words in the 1953 "Chance for Peace" address, Kennedy said, "[N]either can two great and powerful groups of nations take comfort from our present course—both sides overburdened by the cost of modern weapons, both rightly alarmed by the steady spread of the deadly atom, yet both racing to alter that balance of terror that stays the hand of mankind's final war." This called for negotiation, Churchill's "jaw-jaw" over "war-war." Characteristically, Kennedy would pose that challenge as a collective one, something for "us" as Americans. "So let us begin anew," he invited his compatriots, "remembering on both sides that civility is not a sign of weakness, and sincerity is always subject to proof. Let us never negotiate out of fear. But let us never fear to negotiate."

And here was another famous Kennedy locution: *antimetabole,* a word derived from the Greek and meaning the repetition of words in transposed order. Kennedy and Sorensen loved the device ("Ask not what your country can do for you—ask what you can do for your country"). It expressed well Kennedy's sense of irony, complexity, and play, and it conveyed rhetorically the powerful idea of human choice: Would we choose to negotiate or be overwhelmed by fear?

Again and again, Kennedy would call on "us" to move forward, sometimes "us" Americans, and sometimes "us" meaning both the United States and Soviet Union. "Let both sides explore what problems unite us instead of belaboring those problems which divide us." After years of false steps, "let both sides, for the first time, formulate serious and precise proposals for the inspection and control of arms." "Together let us explore the stars, conquer the deserts, eradicate disease, tap the ocean depths, and encourage the arts and commerce."

In the inaugural address, Kennedy also emphasized, as Churchill and Eisenhower had before him, that progress would be incremental, not a single transformation. He aimed for a "beachhead of cooperation to push back the jungle of suspicion." The quest for peace would take time. It would not be finished in his time or ours, but we must take the first steps:

> All this will not be finished in the first 100 days. Nor will it be finished in the first 1,000 days, nor in the life of this Administration, nor even perhaps in our lifetime on this planet. But let us begin.

The first 100 days would bring the Bay of Pigs. The first 600 days would bring the world to the brink of annihilation. But the first 1,000 days would indeed be time enough to move the world.

Kennedy's Address to the
UN General Assembly (September 25, 1961)

To move the world, Kennedy knew, he would need the world, or most of it, to support the cause of peace. World opinion would sway the opinions of the leading antagonists; a worldwide call for peace would help to enforce any bilateral agreement. So Kennedy's next important speech for peace was eight months after his inaugural, at the opening of the UN General Assembly in September 1961.[17] Here, in front of world leaders, he again emphasized the unprecedented reality of the nuclear age: the ability of man to end human life. The very meaning of war had therefore changed:

> [W]ar appeals no longer as a rational alternative. Unconditional war can no longer lead to unconditional victory. It can no longer serve to settle disputes. It can no longer concern the great powers alone. For a nuclear disaster, spread by wind and water and fear, could well engulf the great and the small, the rich and the poor, the committed and the uncommitted alike. Mankind must put an end to war—or war will put an end to mankind.

Here again was *antimetabole,* conveying humanity's most fundamental choice: to end war or be ended by it.

Kennedy then invoked the grim prospect of threatened annihilation:

> Today, every inhabitant of this planet must contemplate the day when this planet may no longer be habitable.

Every man, woman and child lives under a nuclear
sword of Damocles, hanging by the slenderest of threads,
capable of being cut at any moment by accident or miscalculation or madness. The weapons of war must be
abolished before they abolish us.

He underscored the paradox of the prisoner's dilemma: that
self-interested and supposedly rational behavior leaves all sides at
grave risk, "for in a spiraling arms race, a nation's security may
well be shrinking even as its arms increase."

He then laid out the framework to break the logic of the arms
race. The starting point is to recognize the scope for mutual gain,
to see that this is not a zero-sum struggle and that a negotiated
outcome can benefit all parties. On that basis he called for an approach to disarmament that "would be so far-reaching yet realistic, so mutually balanced and beneficial, that it could be accepted
by every nation." Yet he also recognized that this agreement would
not be reached in one fell swoop. Peace would be a process, which
he dubbed a "peace race" instead of an arms race. The two sides
should "advance together step by step, stage by stage, until general
and complete disarmament has been achieved." This echoed
Churchill and Eisenhower, especially Churchill's rejection of the
assumption that nothing can be settled until everything is settled.
In short, Kennedy urged, "general and complete disarmament
must no longer be a slogan, used to resist the first steps."

Kennedy took pains to underscore what arms control would
not do. His rhetoric, powerfully balancing vision and realism,
gained credibility as he reminded the UN General Assembly what
arms treaties would not accomplish: "Such a plan would not bring
a world free from conflict and greed—but it would bring a world
free from the terrors of mass destruction." Kennedy would use the
same rhetorical tactic later on when he presented the test ban
treaty to the American people by listing all that it wouldn't do,
which served to strongly highlight what it *would* do.

From there, Kennedy proposed a path to disarmament, a six-stage trajectory to success:

First, signing the test-ban treaty by all nations.

Second, stopping the production of fissionable materials for use in weapons, and preventing their transfer to any nation now lacking nuclear weapons.

Third, prohibiting the transfer of control over nuclear weapons to states that do not own them.

Fourth, keeping nuclear weapons from seeding new battlegrounds in outer space.

Fifth, gradually destroying existing nuclear weapons and converting their materials to peaceful uses; and

Sixth, halting the unlimited testing and production of strategic nuclear delivery systems, and gradually destroying them as well.

Despite the wild swings of geopolitics, the brinksmanship of the Cuban Missile Crisis, Kennedy's assassination and Khrushchev's ouster, and the Vietnam War debacle, Kennedy started a process that turned a significant part of his vision on that day at the UN General Assembly into global reality.

The Papal Encyclical
Peace on Earth (April 11, 1963)

Kennedy's Peace Speech would put peace in moral terms, as a basic human right and a reflection of our common humanity. These were powerful and brave sentiments at the height of the Cold War, when readiness for war, not peace, was seen as the measure of strength and patriotism. Kennedy was most likely encouraged to put the moral arguments so directly and cogently in part by Pope John XXIII's encyclical on peace, which was delivered just

weeks before the pope's death from cancer. As the pope said to
Norman Cousins: "World peace is mankind's greatest need. I am
old but I will do what I can in the time I have."[18] The pope also
reflected on the power of speaking about peace:

> We must always try to speak to the good in people.
> Nothing can be lost by trying. Everything can be lost if
> men do not find some way to work together to save the
> peace. I am not afraid to talk to anyone about peace on
> Earth.[19]

The key to the encyclical (which Cousins hand-delivered to
Khrushchev in Russian translation) is that peace is part of a moral
order. Morality demands peace, and peace makes morality possible. Society itself is grounded in moral law, wherein "men recognize and observe their mutual rights and duties."[20] On the global
level, deterrence through arms is no basis for peace. The principle
of deterrence "must be replaced by another, which declares that
the true and solid peace of nations consists not in equality of arms
but in mutual trust alone."[21] Leaders should study the problem
of peaceful adjustment among nations "until they find that point
of agreement from which it will be possible to commence to go
forward towards accords that will be sincere, lasting, and fruitful."[22]

The pope gave hope to the possibility of a negotiated peace:

> We grant indeed that this conviction is chiefly based on
> the terrible destructive force of modern weapons and a
> fear of the calamities and frightful destruction which
> such weapons would cause. Therefore, in an age such as
> ours which prides itself on its atomic energy it is con-
> trary to reason to hold that war is now a suitable way to
> restore rights which have been violated.

Nevertheless, unfortunately, the law of fear still reigns among peoples, and it forces them to spend fabulous sums for armaments, not for aggression they affirm—and there is no reason for not believing them—but to dissuade others from aggression.

There is reason to hope, however, that by meeting and negotiating, men may come to discover better the bonds that unite them together, deriving from the human nature which they have in common; and that they may also come to discover that one of the most profound requirements of their common nature is this: that between them and their respective peoples it is not fear which should reign but love, a love which tends to express itself in a collaboration that is loyal, manifold in form and productive of many benefits.[23]

Eight weeks after issuing this encyclical, Pope John XXIII was dead. Yet his message lived on in Kennedy's Peace Speech, delivered the following week.

Chapter 5.

THE PEACE SPEECH

THE PEACE SPEECH IS a work of magnificent culmination. Kennedy's oratory, backed by Sorensen's gifted phrases, was
always powerful, but never more so than in the Peace Speech,
where rhetoric, history, leadership, and morality converged. In
fact, the two greatest moral themes of the time converged at that
very moment: global peace and human rights. Kennedy's Peace
Speech on Monday, June 10, was followed the very next evening by
his great speech on civil rights. In both cases, Kennedy was not
only the nation's political leader but also its moral leader. In these
speeches Kennedy emphasized the strong interconnectedness of
the global and national agendas, and the inevitable intertwining
of peace and justice.

To this day, Kennedy's speech stands out as a unique approach
to global affairs. Its power derives not only from its bold vision of
peace between the United States and the Soviet Union, but also
from its call upon Americans to reexamine their own attitudes
toward peace. Kennedy's point was basic, yet remarkably unusual

in international affairs: that there was humanity, decency, and valor on both sides of the Cold War divide. And because both sides shared in the same human drama, both sides would share in the gains from peace. A peace agreement was therefore feasible, because it would be mutually beneficial.

Kennedy knew that he needed the American people on his side. Even if he signed a treaty with Khrushchev, it would count for nothing if it died in the Senate. The tragedy of Woodrow Wilson, who succeeded in negotiating the League of Nations only to fail to win Senate ratification for it, was always present in Kennedy's mind. Nor could Kennedy ignore Chamberlain's disaster at Munich, and Kennedy's own vulnerability to right-wing cries of appeasement. Kennedy would need a campaign, state by state, Senate vote by Senate vote, to assure the American people that the treaty was in their interest. His success or failure in the peace campaign, he knew, would augur his success or failure in his reelection campaign the following year. And as a campaign begins with a kickoff speech, the campaign for global peace would begin with one as well. There could be no better venue than the hometown university on commencement day.

At American University

Kennedy had planned on making a major peace speech for months. Ted Sorensen recalled, "The President considered in the early spring of 1963 the idea of delivering a speech on peace, a speech which emphasized our peaceful posture and desires, a speech which talked in terms of a peace race instead of an arms race much as the President's speech to the United Nations in 1961 had done."[1] Norman Cousins, the informal emissary among Kennedy, Khrushchev, and Pope John XXIII, after meeting with Khrushchev urged Kennedy to take a "breathtaking new approach toward the Russian people, calling for an end to the cold war and a fresh start in American-Russian relationships."[2] The American University commencement seemed the perfect time and place.

The speech was prepared by a tight circle, lest a more skeptical administration member try to derail it or water it down. Ted Sorensen worked on the draft with a few key advisers, including McGeorge Bundy, Carl Kaysen, and William Foster, the director of the Arms Control and Disarmament Agency.

Secretary of State Dean Rusk, Secretary of Defense Robert McNamara, and Ambassador Llewellyn Thompson saw it with less than a week to go.* Maxwell Taylor, chairman of the Joint Chiefs of Staff,† and Glenn Seaborg, chairman of the Atomic Energy Commission, were shown the sections regarding nuclear testing just a few days before.[3] Sorensen's draft sailed through with minor amendments. The final drafts showed improvements in phrasing, but no major changes in substance.

Kennedy was in Hawaii the night before the speech for a meeting of the U.S. Conference of Mayors. William Foster recalled the rush to finish: "We worked like hell all day. Then Ted Sorensen, I think, sat up all night with his remarkable ability to polish and write and was able to send each of us and the President the final draft about six or seven in the morning to see if there were changes

---

* Carl Kaysen recalled that tracking all the men down to show them the speech "turned out to be a little bit of an operation," as Thompson was in San Francisco and McNamara was at the Williams College commencement in Williamstown, Massachusetts. He remembered them all reacting to the speech positively. Carl Kaysen, recorded interview by Joseph E. O'Connor, July 15, 1966, 111–112, John F. Kennedy Library Oral History Program.

† Taylor specifically suggested that the draft not be circulated further among the other chiefs. Carl Kaysen recalled, "[P]ersonally he thought it was a good decision but he felt that officially he shouldn't have any comment on it because it was a political decision, it was the President's decision to make; that he thought unnecessary and perhaps unwise, although I'm not sure those were his exact words; to show the draft to his colleagues or the chiefs, that their comments were predictable and he felt no purpose could be served." Kaysen, recorded interview by O'Connor, 113.

Kennedy delivers the Peace Speech (June 10, 1963).

to be made. We had another meeting just before the speech, after
we got the President's comments back by cable."[4] Sorensen flew to
Hawaii to bring the final draft and return with Kennedy on a Sun-
day night flight, during which Kennedy put his final touches on
the address.* Carl Kaysen called in the final suggested changes
from cabinet secretaries.

Upon landing, Kennedy went briefly to the White House and
then straight to the American University campus. The day was
sunny and children played on the grass while college students
awaited their diplomas. President Kennedy rose to the dais to accept an honorary degree and deliver the commencement address.[5]
He saluted the university leaders on the dais together with him, as

---

* Averell Harriman, former ambassador to the Soviet Union, was also on the plane, and
Sorensen recalled that "he liked it very much, which encouraged the President not to
change it further." Theodore C. Sorensen, recorded interview by Carl Kaysen, April 15,
1964, 72, John F. Kennedy Library Oral History Program.

well as his former Senate colleague Robert Byrd, an alumnus of American University. Senator Byrd, he quipped, "earned his degree through many years of attending night law school, while I am earning mine in the next 30 minutes."

Kennedy quickly set the theme of personal responsibility by noting that President Woodrow Wilson had said that "every man sent out from a university should be a man of his nation as well as a man of his time." He expressed confidence that the graduates would offer "a high measure of public service." He quoted the English poet John Masefield, who extolled the university as "a place where those who hate ignorance may strive to know, where those who perceive truth may strive to make others see." That was Kennedy's task that morning. He would use the occasion to discuss a topic "on which ignorance too often abounds," a topic he declared to be "the most important topic on earth: peace." He would make the case that peace with the Soviet Union was both possible and necessary, despite the pervasive fatalism that war was inevitable.

Kennedy defined the challenge in global rather than national terms, the pattern he would follow throughout the twenty-six-minute speech:

> [W]hat kind of a peace do we seek? Not a Pax Americana enforced on the world by American weapons of war. Not the peace of the grave or the security of the slave. I am talking about genuine peace, the kind of peace that makes life on earth worth living, and the kind that enables men and nations to grow, and to hope, and build a better life for their children—not merely peace for Americans but peace for all men and women, not merely peace in our time but peace in all time.

Here is that echo of Churchill, who had sought peace "not only for our time, but for a century to come." We also see both men's deliberate contrast with Neville Chamberlain's appeasement at Munich.

Next, by explaining the logic of the prisoner's dilemma, how two sides can get trapped in a wasteful and dangerous arms race with both ending up the losers, Kennedy showed how peace was possible in a world where war seemed nearly inevitable. First, he had to dispel the idea that a nuclear war could be fought and won:

> Total war makes no sense in an age where great powers can maintain large and relatively invulnerable nuclear forces and refuse to surrender without resort to those forces. It makes no sense in an age where a single nuclear weapon contains almost ten times the explosive force delivered by all the allied air forces in the Second World War. It makes no sense in an age when the deadly poisons produced by a nuclear exchange would be carried by wind and water and soil and seed to the far corners of the globe and to generations yet unborn.

Kennedy then acknowledged the perverse logic of deterrence:

> Today the expenditure of billions of dollars every year on weapons acquired for the purpose of making sure we never need them is essential to the keeping of peace.

Yet he rejected the idea that we should be satisfied or comforted by such a situation:

> But surely the acquisition of such idle stockpiles—which can only destroy and never create—is not the only, much less the most efficient, means of assuring peace.

For Kennedy knew the arms race was not only hugely costly, but an invitation to a devastating blunder, as had nearly occurred just eight months earlier.

Kennedy therefore offered the first of three definitions of peace:

"the necessary, rational end of rational men." Men and women seek security for themselves and their children. Nations must do the same on their behalf. Vast stockpiles of arms in a balance of terror can never deliver the desired security, at least not in comparison with peace itself. But can peace really be achieved, or is that merely an illusion, a way to be suckered and overtaken by the other side?

Kennedy's answer was that peace is possible despite the many prophets of doom. The barriers are not only in our adversaries, but also, remarkably and paradoxically, in ourselves:

> Some say that it is useless to speak of peace or world law or world disarmament, and that it will be useless until the leaders of the Soviet Union adopt a more enlightened attitude. I hope they do. I believe we can help them do it. But I also believe that we must reexamine our own attitudes, as individuals and as a Nation, for our attitude is as essential as theirs.

He then bid his countrymen to reexamine their attitudes in four areas: the possibility of peace; the Soviet Union; the Cold War; and freedom at home in the United States.

Our Attitude Toward Peace

Kennedy's first task was to explain why peace should even be considered possible after eighteen years of continuous crisis, following six years of devastating war. He began by raising our hopes: that as humans we can solve even our toughest problems.

> First examine our attitude towards peace itself. Too many of us think it is impossible. Too many think it is unreal. But that is a dangerous, defeatist belief. It leads

to the conclusion that war is inevitable, that mankind is doomed, that we are gripped by forces we cannot control. We need not accept that view. Our problems are manmade; therefore, they can be solved by man. And man can be as big as he wants. No problem of human destiny is beyond human beings. Man's reason and spirit have often solved the seemingly unsolvable, and we believe they can do it again.

Yet Kennedy was ever the realist. He quickly cautioned:

I am not referring to the absolute, infinite concept of universal peace and good will of which some fantasies and fanatics dream. I do not deny the value of hopes and dreams but we merely invite discouragement and incredulity by making that our only and immediate goal.

Kennedy invoked his long-held belief that peace would have to be built step by step:

Let us focus instead on a more practical, more attainable peace, based not on a sudden revolution in human nature but on a gradual evolution in human institutions—on a series of concrete actions and effective agreements which are in the interest of all concerned. There is no single, simple key to this peace; no grand or magic formula to be adopted by one or two powers. Genuine peace must be the product of many nations, the sum of many acts. It must be dynamic, not static, changing to meet the challenge of each new generation.

Kennedy thereby arrived at his second definition: "For peace is a process—a way of solving problems."

But how can peace be reached with such an implacable foe as the Soviet Union? Begin, said Kennedy, with a realistic assessment of the conditions for peace:

> World peace, like community peace, does not require that each man love his neighbor, it requires only that they live together in mutual tolerance, submitting their disputes to a just and peaceful settlement.

Moreover, echoing the historian and theorist B. H. Liddell Hart, Kennedy reminded Americans that:

> history teaches us that enmities between nations, as between individuals, do not last forever. However fixed our likes and dislikes may seem, the tide of time and events will often bring surprising changes in the relations between nations and neighbors.

"So let us persevere," said Kennedy.

Balancing idealism and practicality, the grand vision of peace with the specific steps to get there, Kennedy charted the way forward with a lesson in good management that can serve a thousand purposes:

> By defining our goal more clearly, by making it seem more manageable and less remote, we can help all people to see it, to draw hope from it, and to move irresistibly towards it.

Here, in one sentence, is the art of great leadership. Define a goal clearly. Explain how it can be achieved in manageable steps. Help others to share the goal—in part through great oratory. Their hopes will move them "irresistibly" toward the goal.

Our Attitude Toward the Soviet Union

U.S. foreign policy speeches from 1945 until the Peace Speech contained a litany of sins committed by the Soviet Union, matched by proclamations of America's unerring and unswerving goodwill. Kennedy sought to make a very different point. He was not interested in condemning the Soviet Union, in "piling up debating points," as he put it later in the speech, but rather in convincing Americans that the Soviet Union shared America's interests in peace, and so could be a partner in peace.

Kennedy began this next section of the speech with a passing critique of Soviet propaganda, dryly commenting on it by quoting the scriptures: "The wicked flee when no man pursueth." (Secretary of State Dean Acheson had used the same biblical reference in 1949 congressional testimony about Soviet opposition to NATO.) But Kennedy quickly turned the tables. He did not want to castigate the Soviet Union but to warn Americans "not to fall into the same trap as the Soviets, not to see only a distorted and desperate view of the other side, not to see conflict as inevitable, accommodation as impossible, and communication as nothing more than an exchange of threats."

For Kennedy's real intention was to humanize the "other side," to show Americans the Soviet interests in peace. He started by reminding Americans not to demonize the Soviet people, however much Americans might abhor the communist system:

No government or social system is so evil that its people must be considered as lacking in virtue. As Americans, we find communism profoundly repugnant as a negation of personal freedom and dignity. But we can still hail the Russian people for their many achievements in science and space, in economic and industrial growth, in culture, in acts of courage.

Peace, Kennedy was emphasizing, requires respect of the other party, a fair and generous appraisal of the other's interests and worth. And Kennedy's praise for the Russians was generous, speaking of their virtue and courage, the classical ideals of citizenship he held highest.

Here, too, he followed Churchill, who had told the American people in "Sinews of Peace":

> There is deep sympathy and goodwill in Britain—and I doubt not here also—towards the peoples of all the Russias and a resolve to persevere through many differences and rebuffs in establishing lasting friendships . . . Above all, we welcome, or should welcome, constant, frequent and growing contacts between the Russian people and our own people on both sides of the Atlantic.

Measure by measure, phrase by phrase, Kennedy brilliantly drew America and the Soviet Union into a shared embrace of peace:

> Among the many traits the peoples of our two countries have in common, none is stronger than our mutual abhorrence of war. Almost unique among the major world powers, we have never been at war with each other.

Here is a paradox indeed. Two countries at the brink of war, yet in their long history, "almost unique among the major world powers, we have never been at war with each other." And Kennedy reminded his listeners of something equally fundamental: the Soviet Union's unmatched sacrifices as the recent ally of the United States in the war against Hitler:

> And no nation in the history of battle ever suffered more than the Soviet Union in the Second World War. At least 20 million lost their lives. Countless millions of homes

and families were burned or sacked. A third of the nation's territory, including two thirds of its industrial base, was turned into a wasteland—a loss equivalent to the destruction of this country east of Chicago.

But what binds the United States and the Soviet Union most in the quest for peace is an irony even stronger than recent history. Though the two countries are the world's strongest, they are also perversely the world's most vulnerable. Such is the reality of the nuclear age:

Today, should total war ever break out again—no matter how—our two countries will be the primary target. It is an ironic but accurate fact that the two strongest powers are the two in the most danger of devastation. All we have built, all we have worked for, would be destroyed in the first 24 hours.

Kennedy did not stop there, with the devastating tally of a future war, but went on to remind Americans (and Russians) of the crushing costs of the current Cold War itself:

And even in the cold war, which brings burdens and dangers to so many countries, including this Nation's closest allies, our two countries bear the heaviest burdens. For we are both devoting massive sums of money to weapons that could be better devoted to combat ignorance, poverty, and disease. We are both caught up in a vicious and dangerous cycle, with suspicion on one side breeding suspicion on the other, and new weapons begetting counter-weapons.

Putting together the pieces, the point is clear and overwhelming. Both the United States and the Soviet Union abhor war. They have never fought each other. They were allies in war. They can admire each other's virtue and valor. They risk mutual annihilation. They are squandering their wealth in an arms race. Therefore, they also share a common interest in peace:

> In short, both the United States and its allies, and the Soviet Union and its allies, have a mutually deep interest in a just and genuine peace and in halting the arms race. Agreements to this end are in the interests of the Soviet Union as well as ours. And even the most hostile nations can be relied upon to accept and keep those treaty obligations, and only those treaty obligations, which are in their own interest.

This last sentence, regarding a country's keeping treaty obligations that are in its own interest, was vintage Churchill, who told the House of Commons in November 1953:

> The only really sure guide to the actions of mighty nations and powerful Governments is a correct estimate of what are and what they consider to be their own interests. I do not find it unreasonable or dangerous to conclude that internal prosperity rather than external conquest is not only the deep desire of the Russian peoples, but also the long-term interest of their rulers.[6]

In reaching this conclusion, Kennedy's rhetoric soared with empathy and insight, in what to my mind are the most eloquent and important words of the speech, and perhaps of his presidency:

> So let us not be blind to our differences, but let us also direct attention to our common interests and the means by which those differences can be resolved. And if we cannot end now our differences, at least we can help

make the world safe for diversity. For in the final analysis, our most basic common link is that we all inhabit
this small planet. We all breathe the same air. We all
cherish our children's futures. And we are all mortal.

Our Attitude Toward the Cold War

Kennedy was not yet done batting down the preconceptions, stereo-
types, and myths that held the world at the brink of the abyss. He
enjoined us to learn the lessons of the Cold War and the harrow-
ing Cuban Missile Crisis. As he had remarked soon after the crisis,
we can't go on living this way. Once again, he returned to a note of
hard realism:

> Third, let us reexamine our attitude towards the cold
> war, remembering we're not engaged in a debate, seek-
> ing to pile up debating points. We are not here distribut-
> ing blame or pointing the finger of judgment. We must
> deal with the world as it is, and not as it might have been
> had the history of the last 18 years been different.

The Cold War can too easily become a hot war. We must com-
port ourselves, on both sides, to avoid disaster. Once again chan-
neling the lessons of Liddell Hart, and of the recent crisis, he
warned us of the dangers of pushing foes to the point of a humiliating retreat:

> And above all, while defending our own vital interests,
> nuclear powers must avert those confrontations which
> bring an adversary to a choice of either a humiliating
> retreat or a nuclear war. To adopt that kind of course in
> the nuclear age would be evidence only of the bank-

ruptcy of our policy—or of a collective death-wish for the world.

Kennedy was teaching us about the dangerous dynamics of crises. These are not just about power, military might, and strategic calculations. They are about pride and humiliation. Any leader must put himself in the position of his counterpart, to understand the implications of his or her own actions for *the other side*—in human, psychological, and social terms.

Kennedy spoke about America's weapons, emphasizing their defensive posture, calling them "nonprovocative, carefully controlled, designed to deter, and capable of selective use." These ideas followed the prevailing doctrine of deterrence, emphasizing an equilibrium in which neither side instills the fear of a first strike. Yet they are probably the least persuasive part of the speech. However the United States may have viewed its military might, the Soviet Union continued to harbor the belief that the United States was preparing a first strike. And this was not mere propaganda; it was a real and palpable fear.

To defuse the tensions of the Cold War, Kennedy had taken steps toward an "increased understanding between the Soviets and ourselves," building on "increased contact and communication." He endorsed a hotline for direct contact between the two sides, having experienced the laborious difficulties of communication during the Cuban Missile Crisis, when every lost minute risked a devastating mistake.

Much more important, he called for a resumption of disarmament talks, implicitly returning to the timetable he had proposed at the UN General Assembly in September 1961:

> Our primary long range interest . . . is general and complete disarmament, designed to take place by stages, permitting parallel political developments to build the new institutions of peace which would take the place of arms.

Kennedy's primary focus in these disarmament talks would be a nuclear test ban treaty:

The only major area of these negotiations where the end is in sight, yet where a fresh start is badly needed, is in a treaty to outlaw nuclear tests. The conclusion of such a treaty, so near and yet so far, would check the spiraling arms race in one of its most dangerous areas. It would place the nuclear powers in a position to deal more effectively with one of the greatest hazards which man faces in 1963, the further spread of nuclear arms. It would increase our security; it would decrease the prospects of war. Surely this goal is sufficiently important to require our steady pursuit, yielding neither to the temptation to give up the whole effort nor the temptation to give up our insistence on vital and responsible safeguards.

Kennedy prioritized a ban on nuclear testing for several reasons: widespread public concern over nuclear fallout from the tests, which had steadily grown since several Japanese fishermen died from fallout poisoning after a U.S. nuclear test in 1954; a hope that a test ban would slow proliferation, notably to China; a belief among scientists that weapons design could proceed even without the tests; and an overarching hope that an agreement on tests would create the momentum for further agreements.

Kennedy concluded the Peace Speech with two important announcements. The first was that Khrushchev, Kennedy, and U.K. prime minister Harold Macmillan, the leaders of the three nuclear powers, had just agreed to talks in Moscow to try to complete a test ban treaty. The second was a matter of goodwill, that the United States would not conduct nuclear tests as long as other states did not do so. The United States, said Kennedy, will not be the first to resume testing. This was a signal that the United States would renew a cooperative strategy, and would stay cooperative as long as its counterparts did as well. Both announcements were interrupted by the vigorous applause of those gathered: the listeners that morning recognized that something new and important was getting under way.

Freedom at Home

The great domestic struggle of 1963 was the heating up of the civil rights movement. The day after the Peace Speech was the first day of racial integration at the University of Alabama.* Kennedy had flown to Hawaii just before the Peace Speech in order to address the nation's mayors on the subject of the civil rights crisis. And he would address it again in a televised talk to the nation the day after the American University speech.† Presidents don't have the luxury of confronting one great issue at a time; for Kennedy, they came simultaneously.

Moreover, Kennedy understood that the crises at home and abroad were not separate but intertwined. Indeed, the Soviet Union had effectively used American racial tensions to embarrass the United States and accuse it of hypocrisy in regards to human rights. In Kennedy's view, each crisis was a test of the nation's valor in its quest for peace and freedom. "The quality of our own society must justify and support our efforts abroad." He termed it an age-old faith that "peace and freedom walk together." At American University, Kennedy spoke obliquely about the civil rights move-

---

* Under the 1954 Supreme Court decision *Brown v. Board of Education,* public educational facilities could not be segregated. When two African-American students were accepted to the University of Alabama, Governor George Wallace, who had campaigned on the promise of "segregation now, segregation tomorrow, segregation forever," physically blocked the doorway to prevent the students from registering. This crisis of both civil rights and federal-state government relations prompted President Kennedy to federalize the Alabama National Guard to protect the students and ensure they could register at the university.

† In a sad irony, the day after Kennedy's civil rights address and two days after the Peace Speech, civil rights activist Medgar Evers was assassinated in his own driveway.

ment, including the spring Birmingham campaign of civil rights demonstrations led by Martin Luther King Jr. and the unrest around the desegregation of the University of Alabama. Not mentioning the drama itself, Kennedy noted, "In too many of our cities today, the peace is not secure because freedom is incomplete." In addition to the responsibilities of government, he declared it the responsibility of every citizen "in all sections of this country to respect the rights of others and respect the law of the land." This line elicited the third rousing applause of the day; clearly the drama in Alabama was very much on the public's mind.

Peace as a Human Right

Kennedy had first defined peace in strategic terms: as the necessary rational end of rational men. He had then defined peace in political terms: as a way of solving problems. He concluded by defining peace in human terms: as a birthright, a human right alongside life, liberty, and the pursuit of happiness. Once more, Kennedy articulated this notion in the most eloquent way:

> And is not peace, in the last analysis, basically a matter of human rights: the right to live out our lives without fear of devastation; the right to breathe air as nature provided it; the right of future generations to a healthy existence?

In honoring this right, Kennedy went on, humanity honors a higher cause. "When a man's way[s] please the Lord," Kennedy said, quoting scripture, "He maketh even his enemies to be at peace with him." This recalled the closing words of his inaugural address, where he asked for God's blessing and help, "but knowing that here on Earth God's work must truly be our own."

Kennedy closed the speech with one final message of peace for the Soviet Union. Soviet leaders had long feared that the U.S. intended to launch a first nuclear strike, and they viewed the network of U.S. military bases in Europe and Asia as the staging grounds for that attack. Kennedy therefore reassured them that the United States "will never start a war":

> We do not want a war. We do not now expect a war. This generation of Americans has already had enough— more than enough—of war and hate and oppression. We shall be prepared if others wish it. We shall be alert to try to stop it. But we shall also do our part to build a world of peace where the weak are safe and the strong are just. We are not helpless before that task or hopeless of its success. Confident and unafraid, we must labor on—not towards a strategy of annihilation but towards a strategy of peace.

As Clear as the Scriptures

The fast-moving events in Alabama gave Kennedy no moment to savor his new peace initiative, no moment to rest. One day later he was back at the lectern, this time at the White House, addressing the American people on the bitterly contested issue of civil rights.[7]

Just as in the Peace Speech, Kennedy appealed to conscience and reflection. Just as he had bade Americans to consider their attitude toward peace, the next day he bade them to consider their attitude toward race. In recounting that the Alabama National Guard had been required that day to integrate the University of Alabama, Kennedy asked every citizen to "stop and examine his conscience about this and other related incidents."

Kennedy returned to basic human rights, again in the context of the worldwide struggle for freedom:

This Nation was founded by men of many nations and backgrounds. It was founded on the principle that all men are created equal, and that the rights of every man are diminished when the rights of one man are threatened.

Today we are committed to a worldwide struggle to promote and protect the rights of all who wish to be free. And when Americans are sent to Vietnam or West Berlin, we do not ask for whites only. It ought to be possible, therefore, for American students of any color to attend any public institution they select without having to be backed up by troops.

In yet another phrase that permanently entered the American political lexicon, Kennedy proclaimed, "We are confronted primarily with a moral issue. It is as old as the scriptures and is as clear as the American Constitution." The heart of the question, said Kennedy, was this:

If an American, because his skin is dark, cannot eat lunch in a restaurant open to the public, if he cannot send his children to the best public school available, if he cannot vote for the public officials who represent him, if, in short, he cannot enjoy the full and free life which all of us want, then who among us would be content to have the color of his skin changed and stand in his place?

Who among us would then be content with the counsels of patience and delay?

Kennedy announced that he would submit a Civil Rights Act to the Congress—legislation that would ultimately be passed in the year after his death thanks to the extraordinary and courageous efforts of his Texan vice president and successor, Lyndon B. Johnson.

In the course of these two days, with these two speeches, Kennedy crossed the threshold from charming, skilled politician to moral leader. In just two days, Kennedy had charted a political course, set a strategy, grasped the nettle of peace, and backed the civil rights movement with the force of the executive branch of the federal government. Soon he was off to Europe to continue the campaign for peace, this time among the allies.

Chapter 6.

THE CAMPAIGN FOR PEACE

THE AMERICAN UNIVERSITY SPEECH launched Kennedy's campaign for peace, which would continue intensively for 100 days, through negotiations in Moscow, ratification of the test ban treaty by the U.S. Senate, and presentation to the world community at the United Nations on September 20. Kennedy's efforts were relentless and carefully targeted. He traveled to Europe to bolster the Western alliance; spoke to the American people; worked the Congress; and continually consulted world leaders. Cajoling, convincing, and always charming, his words were powerful instruments of persuasion.

Public Reaction to the Speech

The U.S. reaction to the Peace Speech was positive, though muted. Too many hopes in the past had been thwarted. Too many negotiations had come to naught, and most Americans blamed the Soviet Union for this. Yet pundits and the public knew that Kennedy had done something new, presenting the difficulties of the Cold War in a novel light. A few recognized the speech as a historic shift, though only the success of the forthcoming negotiations could confirm the speech's significance. The final verdict would have to wait.

The Washington Post editorialized that the speech "was much more than an appeal for a ban on nuclear testing. It was, indeed, another bid for an end to the cold war."[1] But the paper cautioned: "Many similar gestures on the part of former President Eisenhower as well as President Kennedy have brought only meager responses from the Communist bloc." The announcement of new test ban negotiations "must be read in the light of many failures of similar attempts in the past."

The New York Times noted that the "most important plea in President Kennedy's eloquent address" was his call to "re-examine our attitude toward the Soviet Union."[2] The editorial concluded with a strong endorsement of Kennedy's attempt at peace. "The search must be to find a truce so the world can live in peace while the arms race is halted. If the meeting in Moscow makes a start in that direction it will be a great moment for mankind."

Walter Lippmann, the most redoubtable political commentator of the era, gave his support as well:

> We on our part and the Russians on their part have raised higher than the iron curtain an impenetrable fog of suspicion . . . The President's address is more than a talk. It is a wise and shrewd action intended primarily to improve the climate of East-West relations.[3]

Lippmann rightly pointed out that for Kennedy and Khrushchev, the idea that one side can "bury the other" (as Khrushchev had famously proclaimed) had become "nonsensical." "In the age of nuclear parity," he wrote, echoing both Kennedy and Khrushchev, "there is no alternative to coexistence."

There was also plenty of critical U.S. press. Columnist Roscoe Drummond of *The Christian Science Monitor*, for example, dismissed Kennedy's hope of making the world safe for diversity. "This is not the Soviet objective," he wrote. "Undoubtedly the Kremlin wants to avoid nuclear war, but short of nuclear war to make the world unsafe for diversity." There was little to do to ease tensions because "the Soviet determination to impose its political will and economy system on others" was "basic Communist doctrine."[4]

The international press reaction was almost universally positive, with two notable and predictable exceptions: China and France. Both planned to build their own nuclear arsenals, as they viewed nuclear weapons as the ultimate guarantor of national power and independence, and did not trust their nuclear-armed allies to defend them. Indeed, by 1963, the Soviet Union and China were essentially antagonists rather than allies. China, in addition to its own nuclear aspirations, had long harangued the Soviet Union for any hints of rapprochement or détente with the United States. The French press expressed skepticism that any effective agreement would emerge from the coming negotiations. The U.S. Information Agency report on international reactions to the speech (coincidentally written by USIA deputy director Thomas Sorensen, Ted's brother) noted that "[s]ome [French] papers suggested that both Macmillan and the President need a foreign success for domestic reasons, and thus may be prepared to make some concessions—especially at France's expense. They also emphasized that a test ban treaty would not bind France or Communist China."[5] For the rest of the world, though, the speech raised hopes, albeit hopes tempered by the long history of diplomatic dead ends.

The greatest enthusiasm was in the United Kingdom, which was to be the third signatory of the pending treaty. Among the allies, Prime Minister Macmillan had been the strongest and most persistent proponent of reopening negotiations, and he had worked closely with Kennedy to achieve it. The *Daily Mail* called the American University address Kennedy's "greatest speech." The *Evening Standard* noted a "political climate that is more refreshing and hopeful than for a very long time."[6] *The Times* of London praised Kennedy's "particular contribution" of emphasizing "the need to respect each other's interests, to accept honest differences, and to refrain from imposing alien systems on smaller countries."[7] In *The Guardian,* the British parliamentarian Richard Crossman called the speech "the most significant American policy declaration for many years," adding, "The more carefully one reads the speech, the more clear it becomes that the President has at last found the courage to call off the cold war."[8]

Soviet Reactions

Soviet reactions would be most vital to success. The White House carefully monitored the early Soviet reactions via U.S. government cables sent from Moscow. The news was encouraging. First, the two major newspapers *Pravda* and *Izvestiya,* which had a combined circulation of ten million readers, both carried the text of the president's speech in full, a rarity for the Soviet press.[9] Second, the Soviets allowed the Voice of America and the BBC to transmit the speech by radio without the usual jamming of the airwaves. Clearly, the Soviet authorities were intent on making the Soviet people aware of the speech. The media reactions were also generally favorable. One U.S. report cited a Soviet news commentator's statement that "hopes have emerged for a radical improvement of the international climate."[10] The general tenor of the Soviet media was that Kennedy's new policy proposals were a step forward, in line with the long-standing Soviet call for "peaceful coexistence."

Of greatest importance were the reactions of the Soviet leadership. Khrushchev's reception of the speech was just as Kennedy

intended: extremely positive and open to a treaty. British opposition leader Harold Wilson met Khrushchev soon after the
speech, and according to Kennedy's special assistant, the historian
Arthur Schlesinger Jr., Wilson "found [Khrushchev] deeply im-
pressed and considerably more open minded about the test ban."[11]
Khrushchev called in Kennedy's envoy Averell Harriman to tell
him that Kennedy's "was the greatest speech by any American
President since Roosevelt."[12] A CIA report the day after the speech
noted:

> The Soviets were favorably surprised by the tenor of
> President Kennedy's 10 June speech because it reflected
> a broad progressive approach toward solving current
> problems. The atmosphere created by this speech is now
> such that the possibilities of agreeing on a test ban treaty
> are very good. No chief of state would make such a
> speech unless he were completely convinced that agree-
> ment was probable. The only problem in the past which
> prevented a test ban treaty was Soviet doubt of the sin-
> cerity of U.S. intentions to enter into such an agree-
> ment . . . President Kennedy's speech has gone a long
> way toward assuaging Soviet doubts of U.S. sincerity.[13]

In Khrushchev's first public remarks on the speech, made in an
interview with *Pravda* and *Izvestiya* on June 14, he was more mea-
sured and circumspect. He deemed it "a step forward in a realistic
appraisal of the international situation," and one that "stressed
the need of finding ways which would rid mankind of the arms
race and the threat of a thermonuclear world war."[14] Yet he also
signaled many points of disagreement, including inspections,
U.S. overseas bases, and the U.S. suppression of national liberation movements. Khrushchev's hedging was likely connected to
strained relations with the Chinese, who had sent a critical open
letter to the Soviet Communist Party the same day.[15]

While the signals were not perfect, it did seem that the Soviet Union was ready for at least some agreements. To bolster those prospects and to meet with key allies in the midst of these negotiations, Kennedy headed to Europe in late June. His travels took him to raucously enthusiastic public events in Germany and Ireland, and less public stops in England and Rome. The whirlwind trip, from June 23 to July 2, was among the most memorable nine days of Kennedy's presidency, and indeed of his life.

The Trip to Europe

John Kennedy's arrival in Europe in late June in many ways recalled Woodrow Wilson's arrival in Europe forty-five years earlier. In both cases, Europeans looked to the American president for deliverance. Wilson personified Europe's hope for a just and lasting peace after the most devastating war in the continent's long and bloody history. Kennedy embodied the hopes of Western Europe for peace and economic dynamism in the Cold War era. To a war-weary continent, Kennedy offered youth, charm, and American optimism. He no doubt had Wilson very much on his mind for more reasons than their similar receptions. Wilson's triumphant European arrival was bookended by his failure to win Senate confirmation of the Paris peace treaty, which doomed the new League of Nations. Wilson drove himself nearly to death in his frenzied campaign for treaty ratification, suffering an incapacitating stroke in the process.

Wilson had arrived in Europe to adoring masses. "As soon as he set foot on French soil, an explosion of celebrations began. French and American soldiers lined the streets of Brest as the Wilsons rode in an open car under triumphal arches of flowers . . . Hordes of cheering people packed the sidewalks and hung out of every window."[16] Kennedy's greeting was no less euphoric. Schlesinger described Kennedy's entrance into Berlin, with "three-fifths of the population of West Berlin streaming into the streets, clapping, waving, crying, cheering, as if it were the second coming."[17]

Kennedy's itinerary was geared to the treaty negotiations. The first stops were in Germany, first Frankfurt and then Berlin, the front line of the Cold War, with a political leadership that had to be handled with care. West German leaders, especially Chancellor Konrad Adenauer, reacted nervously to any signs of U.S. rapprochement with the Soviet Union, out of fear that West Germany's interests might be undermined in some grand bargain on European security. Whenever Kennedy took pains to emphasize that he would make no agreements at the expense of the Western allies, he had West Germany first in mind.*

Let Them Come to Berlin

The speech in Berlin marked the most remarkable occasion of the whole European trip.[18] After taking a tour of the western part of the city, including a solemn visit to the Berlin Wall, Kennedy's car slowly proceeded through throngs of cheering people, finally arriving in the city center. Kennedy's visit was an unprecedented public event. Over 1,500 journalists were accredited to cover it,† and many businesses and schools were closed for the day.[19] Hundreds of thousands of people filled the square; the empty streets of East Berlin could be seen just over the Berlin Wall. It was an emo-

---

* And the Germans were certainly listening—when Kennedy arrived in West Germany, Adenauer greeted him by quoting the passage from the American University speech stating that the United States would not make concessions at the expense of its allies. Vito N. Silvestri, *Becoming JFK: A Profile in Communication* (Westport, CT: Praeger, 2000), 229.

† Working out of Schoeneberg City Hall, the correspondents in one day went through 4,540 sandwiches, 690 sausages, 987 bottles of beer, 402 packs of cigarettes, and 23 bottles of whiskey. Andreas W. Daum, *Kennedy in Berlin* (Washington, DC: German Historical Institute, 2008), 28.

Kennedy delivers his *"Ich bin ein Berliner"* speech in Rudolph Wilde Platz, Berlin (June 26, 1963).

tional crowd that, in turn, moved Kennedy into a flight of improvised rhetoric that went further than he himself had intended in the prepared draft.

"I am proud to come to this city," he began, as he sized up the mass of humanity before him at ground zero of the Cold War:

> Two thousand years ago the proudest boast was *"civis Romanus sum."* Today, in the world of freedom, the proudest boast is *"Ich bin ein Berliner."*

The huge crowd roared its delight. As Kennedy's locution was not exact (the *ein* was out of place), the translator repeated Kennedy's

line, prompting Kennedy to quip that he appreciated "my interpreter translating my German!"

Kennedy then threw down the gauntlet, as Ronald Reagan
would more than twenty years later when he called on another
Soviet leader to "tear down this Wall." Standing before the grim
wall, in a city divided between West Berlin's rising prosperity and
East Berlin's evident gray stagnation, Kennedy told the assembled
multitude:

> There are many people in the world who really don't un-
> derstand, or say they don't, what is the great issue be-
> tween the free world and the Communist world. Let
> them come to Berlin. There are some who say that com-
> munism is the wave of the future. Let them come to
> Berlin. And there are some who say in Europe and else-
> where we can work with the Communists. Let them
> come to Berlin. And there are even a few who say that it
> is true that communism is an evil system, but it permits
> us to make economic progress. *Lass' sie nach Berlin
> kommen.* Let them come to Berlin.

Kennedy continued, "I know of no town, no city, that has been
besieged for 18 years that still lives with the vitality and the force,
and the hope and the determination of the city of West Berlin." He
bid his listeners to "lift your eyes beyond the dangers of today, to
the hopes of tomorrow":

> Freedom is indivisible, and when one man is enslaved,
> all are not free. When all are free, then we can look for-
> ward to that day when this city will be joined as one
> and this country and this great Continent of Europe in
> a peaceful and hopeful globe. When that day finally
> comes, as it will, the people of West Berlin can take

sober satisfaction in the fact that they were in the front lines for almost two decades.

And then he closed with his reaffirmation that "as a free man, I take pride in the words *'Ich bin ein Berliner!'*"

The crowd was euphoric, and Kennedy was exhilarated. He had achieved his political purpose: to secure the support and confidence of the German people. Still, he knew that he had gone a step too far with his anti-communist rhetoric. Swept away by emotion, his language seemed to contradict the conciliatory message of the Peace Speech. Years later, his close aides Kenny O'Donnell and Dave Powers recalled, "Kennedy's fighting speech in Berlin . . . actually was a grave political risk, and he knew it. Such a heated tribute to West Berlin's resistance to Communism could have undone all the success of his appeal for peace and understanding with the Soviets in his American University speech two weeks earlier."[20] To gain the political room to negotiate the treaty by winning German public opinion, while simultaneously closing the deal with Khrushchev, Kennedy had to tread a fine line.

At his next stop that afternoon at the Free University of Berlin, he dialed down the rhetoric a few notches.[21] There he emphasized "the necessity of great powers working together to preserve the human race, or otherwise we can be destroyed." He called on the Western alliance to remain strong, noting that he had traveled to Europe to bolster the alliance's unity. For only through a strong alliance "will genuine, mutually acceptable proposals to reduce hostility have a chance to succeed."[22]

On that day, Kennedy effectively cemented the Western alliance, a remarkable feat considering that only eighteen years earlier Germany was a Nazi state and Berlin was aflame and in ruins. Kennedy had encouraged the German people to be resolute in the defense of freedom. He emotionally connected them to America and the Western alliance. It was a bravura performance. Despite the heated rhetoric, the Berlin speech did not threaten the momentum toward a U.S.-Soviet treaty.* Most important, Kennedy would use the political capital gained in Berlin to deny the West German government's aspirations for nuclear weapons. So, on the main point of substance, not only Kennedy but Khrushchev too triumphed that day in Berlin.

Ireland and Italy

Kennedy's next stop was more nostalgic and personal: Ireland. Kennedy arrived with two of his sisters, Jean Kennedy Smith and Eunice Kennedy Shriver, and the visit was a mix of sightseeing, public events, and memorable speechmaking, notably at the Irish Parliament, the Dáil.²³ There he again called on Europe to join in the quest for peace, and emphasized that all countries, small as well as large, were needed in this global effort. The latter point was not intended for Irish ears alone.

The blood ties of President Kennedy and his hosts were close. The Parliament was welcoming home one of its own. He was proud, he said, to be the first American president to visit Ireland while in office and the first to address the Parliament. He was also, of course, the first Catholic Irish-American president.

Kennedy had come in part to sing Ireland's praises:

This has never been a rich or powerful country, and yet, since earliest times, its influence on the world has been rich and powerful. No larger nation did more to keep Christianity and Western culture alive in their darkest centuries. No larger nation did more to spark the cause

---

* Khrushchev seems to have accepted the speech as political theater, though he did note in a letter to Prime Minister Macmillan that the Berlin speech seemed to have been delivered by "quite a different person" than the orator of the American University speech. Glenn T. Seaborg, *Kennedy, Khrushchev, and the Test Ban*, ed. Benjamin S. Loeb (Berkeley: University of California Press, 1981), 230.

Kennedy's motorcade drives through Cork, Ireland (June 28, 1963).

of independence in America, indeed, around the world. And no larger nation has ever provided the world with more literary and artistic genius.

Yet he had also come to speak of world peace, and of the role of nations large and small. Quoting John Boyle O'Reilly, a nineteenth-century Irish-American poet who hailed from Boston:

> *The world is large when its weary leagues two loving*
>   *hearts divide,*
> *But the world is small when your enemy is loose on the*
>   *other side.*

Yes, said Kennedy, "[T]he world is even smaller today, though the enemy of John Boyle O'Reilly [the British Empire] is no longer a hostile power" to Ireland. Picking up one of his favorite themes, Kennedy noted that even ancient enmities do not last forever, and that peace is a choice:

Indeed, across the gulfs and barriers that now divide us, we must remember that there are no permanent enemies. Hostility today is a fact, but it is not a ruling law. The supreme reality of our time is our indivisibility as children of God and our common vulnerability on this planet.

Ireland, he said, must play its part, especially in the United Nations: "The major forum for your nation's greater role in world affairs is that of protector of the weak and voice of the small, the United Nations. From Cork to the Congo, from Galway to the Gaza Strip, from this legislative assembly to the United Nations, Ireland is sending its most talented men to do the world's most important work—the work of peace." And, added Kennedy, other small nations should follow this lead:

I speak of these matters today—not because Ireland is unaware of its role—but I think it important that you know that we know what you have done. And I speak to remind the other small nations that they, too, can and must help build a world peace. They, too, as we all are, are dependent on the United Nations for security, for an equal chance to be heard, for progress towards a world made safe for diversity.

Kennedy's peace campaign, in its essence, held that mankind can choose its fate, that we are not "gripped by forces we cannot control," as he had put it at American University. So it was fitting that he gave an Irish spin to that philosophy, his abiding belief that we can shape our future by dreaming of a better world:

George Bernard Shaw, speaking as an Irishman, summed up an approach to life: Other people, he said "see things

and . . . say 'Why?' . . . But I dream things that never
were—and I say: 'Why not?'"

It is that quality of the Irish—that remarkable combination of hope, confidence and imagination—that is
needed more than ever today. The problems of the world
cannot possibly be solved by skeptics or cynics, whose
horizons are limited by the obvious realities. We need
men who can dream of things that never were, and ask
why not.

Kennedy had given Shaw's words an interesting twist. Origi-
nally, in Shaw's play *Back to Methuselah,* the words were spoken by
the Serpent to Eve.[24] In Kennedy's retelling, they were the spirit of
man striving to create his own Eden on Earth. Robert Kennedy
would choose this phrase as the motto of his 1968 presidential
campaign. And, most poignantly, we remember them as the closing words of Ted Kennedy's moving eulogy of his brother Robert
after the latter's assassination.[25]

From Ireland Kennedy went to England. His first stop there was
the most private of all: a heartrending visit to the gravesite of his
beloved younger sister Kathleen (Kick), who had married a young
British aristocrat who was killed in battle after D-Day. She herself
died in a 1948 plane crash at only twenty-eight. From the late 1930s
till her death a decade later, Kick had brought the young John into
the dashing world of youthful British protégés of Churchill, young
men who would continue to play an important role in shaping
Kennedy's worldview during his presidency. Among these was
one of Kennedy's closest confidants in foreign policy, David
Ormsby-Gore, the future British Ambassador to the United States
during Kennedy's presidency.

Next he visited with Prime Minister Macmillan, for one-on-one
meetings to coordinate positions on the upcoming negotiations in
Moscow. Macmillan would instruct his negotiators to follow the
U.S. lead on all points. Finally, from the United Kingdom he went

to Italy. Kennedy traveled first to Rome to meet with the government and with the new pope, Paul VI, following the death of Pope John XXIII on June 3. Then, for his last stop in Europe, Kennedy went to NATO headquarters in Naples. There he put a capstone on his peace initiative, closing the European trip by declaring, "The purpose of our military strength is peace. The purpose of our partnership is peace":

> So our negotiations for an end to nuclear tests and our opposition to nuclear dispersal are fully consistent with our attention to defense—these are all complementary parts of a single strategy for peace. We do not believe that war is unavoidable or that negotiations are inherently undesirable. We do believe that an end to the arms race is in the interest of all and that we can move toward that end with injury to none. In negotiations to achieve peace, as well as preparation to prevent war, the West is united, and no ally will abandon the interests of another to achieve a spurious detente. But, as we arm to parley, we will not reject any path or refuse any proposal without examining its possibilities for peace.[26]

His campaign was paying dividends. As Sorensen related, "That night, as we flew back to Washington, a message radioed to the plane told of a Khrushchev speech that day in East Berlin. It endorsed an atmospheric nuclear test-ban treaty."[27]

The Final Round of
Negotiations in Moscow

The final challenge for achieving a test ban treaty was the question of underground tests and the still-vexing issue of inspections. The question was how to verify and enforce a treaty. For nuclear testing in the air, in space, and underwater, remote sensing of a nuclear test was possible. For example, air samples could be collected to test for radioactive fallout. The challenge was verifying a ban on underground tests. Here, the United States scientific community advising Eisenhower and then Kennedy had long insisted that monitoring underground tests required onsite inspections, while the Soviets argued that seismological detectors could pick up underground tests remotely, and distinguish the signals of a nuclear test from those of a naturally occurring earthquake. They strongly opposed onsite inspections, believing them to be pretexts for espionage.

The standoff over onsite inspections had blocked a test ban treaty for years, and almost did so again in late 1962, when a disagreement erupted over each side's proposals for the number of onsite inspections needed per year. After years of rejecting any onsite inspections, Khrushchev finally agreed to a small number of inspections, three, noting to Kennedy in a letter on December 19 that a United States negotiator had suggested to his Soviet counterpart that the United States would agree to two to four inspections per year.* Yet after the Politburo assented to three inspections, Khrushchev was informed by Kennedy in a letter of December 28 that "there appears to have been some misunderstanding," as the U.S. position was and had been a minimum of eight to ten onsite inspections per year, a number that Khrushchev knew his colleagues would reject.

When Khrushchev read JFK's message that the United States was demanding at least eight inspections rather than the three he believed the United States had agreed to, he exploded. It seemed

---

* In his letter of December 19, Khrushchev wrote, "We noted that on this October 30, in conversation with First Deputy Foreign Minister of the USSR V. V. Kuznetsov in New York, your representative Ambassador Dean stated that, in the opinion of the U.S. Government, it would be sufficient to carry on 2–4 on-site inspections each year on the territory of the Soviet Union" (Khrushchev to Kennedy, December 19, 1962, in *Foreign Relations of the United States, 1961–1963*, volume VI, *Kennedy-Khrushchev Exchanges* [Washington, D.C.: U.S. Government Printing Office, 1996], Document 85).

yet another case of U.S. backtracking.[28] He was already far out on a limb vis-à-vis his Politburo colleagues and the Chinese in the wake of the Soviet reversal in the Cuban Missile Crisis, and knew that he could not go back to the Politburo for a number of inspections greater than three.

When Khrushchev met with Norman Cousins at his Black Sea villa in the spring of 1963, the Soviet leader put the case this way:

> People in the United States seem to think I am a dictator who can put into practice any policy I wish. Not so. I've got to persuade before I can govern. Anyway the Council of Ministers agreed to my urgent recommendation [on three inspections]. Then I notified the United States that I would accept three inspections. Back came the American rejection. They now wanted—not three inspections or even six. They wanted eight. And so once again I was made to look foolish. But I can tell you this: it won't happen again.[29]

One way out of the stalemate, proposed as early as 1959 by the United States but rejected by the Soviets at the time, was a partial test ban, covering the atmosphere, the oceans, and space, where monitoring from afar was possible, but excluding underground tests. Kennedy and Macmillan repeated the proposal for the "three-environment treaty" in September 1961. The Soviets rejected the proposal again, dismissing it as a green light for underground testing, which the Soviet leaders believed would favor the United States. Kennedy and his foreign policy advisers also viewed the partial test ban as inferior to a comprehensive test ban, as the United States worried that underground tests would provide aspiring nuclear nations such as China and France with a route to the bomb.[30] The U.S. military, however, preferred a partial test ban, as it would allow them to continue with their own underground tests.

Such was the state of play into the summer of 1963. But on

July 2, the Soviets acquiesced to a partial test ban, one that would not cover underground tests. Khrushchev put it this way in his speech in East Berlin:

> Having carefully weighed the situation, the Soviet Union, moved by a sense of great responsibility for the fate of the peoples . . . [and] since the Western powers are impeding the conclusion of an agreement on all nuclear tests, expresses its readiness to conclude an agreement on the cessation of nuclear tests in the atmosphere, in outer space, and under water.[31]

Several factors no doubt contributed to this decision. Most important was the desire on both sides to achieve a concrete result after so many years of floundering, and after the shared shock of the Cuban Missile Crisis. Kennedy's peace initiative had done much in a short period to soften Khrushchev's opposition to a partial test ban. The growing rift between the Soviet Union and China added to the impetus for achieving a treaty quickly. Perhaps the partial nuclear test ban would slow China's path to its own nuclear weapon, an increasingly important objective for the Soviet Union, especially as Khrushchev had come to view the Chinese leadership under Mao as provocative, unstable, and dangerous to world peace.

The bottom line was that neither side could bridge the gap regarding onsite inspections. The Soviet resistance to more inspections was real, not bluff. And Kennedy also faced severe political limits on what he could concede on the issue. He was fully aware that the Senate would have to pass any treaty with a two-thirds vote, and that many hardliners in the military, the scientific community, and Congress were against any agreement at all. A high number of onsite inspections was a sine qua non for the hardliners, just as much as it was a red light for the Soviet side.

A week before the negotiations began, the lead U.S. negotiator, Averell Harriman, received his instructions.[32] Try one more time to close the gap on inspections for a comprehensive test ban. If that was not feasible, go for the three-environment agreement. And to the maximum extent possible, unlink the test ban negotiations from other Soviet proposals, such as a non-aggression pact, which would entangle more allies in the negotiations and delay the conclusion of a test ban treaty.

The three-nation negotiations on the test ban treaty got under way in Moscow on July 15.

Foreign Minister Andrei Gromyko was appointed to head the Soviet negotiating team as counterpart to Harriman, signaling the great importance the Soviet leadership attached to the negotiations. Moreover, Khrushchev himself sat through the entire first day. Both sides wanted an agreement, but there was a lot of ground to cover as each side tested the other's boundaries. The Soviets first proposed that the test ban discussions be combined with a non-aggression pact. As planned, Harriman demurred, explaining that this would complicate and delay the negotiations. After much to-ing and fro-ing, the two sides agreed to a paragraph in the communiqué announcing their mutual interest in exploring a non-aggression pact.

The next challenge was defining the scope of the treaty: partial or comprehensive. Harriman probed whether a comprehensive treaty including an underground testing ban might still be considered, but Khrushchev shut the door firmly on further discussions regarding onsite inspection. The only course ahead was a partial treaty, with a clause in the final agreement stipulating that the agreement was "without prejudice to the conclusion of a treaty resulting in the permanent banning of all nuclear test options."

Two other modest hurdles remained. The first was agreement on peaceful uses of nuclear explosions. Khrushchev argued that those, too, should be ruled out, because even ostensibly peaceful uses (such as to mine resources, divert rivers, or carry out other such activities) could be used surreptitiously to gain military information. The United States accepted the point.

The second hurdle was a bit trickier. It concerned the right to withdraw from the treaty, for example in the case of war, or in the event that China or another country carried out its own nuclear tests, thereby pressuring the United States or the Soviet Union to resume testing. A withdrawal clause, Kennedy felt, would be vital to the successful passage of the treaty by the Senate. Khrushchev initially demurred, arguing that the treaty would be devalued by having a built-in aura of impermanence. In the end, after considerable discussion, the two sides agreed on a country's right to withdraw from the treaty "if it decides that extraordinary events, related to the subject matter of this Treaty, have jeopardized the supreme interests of its country." When China detonated its first nuclear device in October 1964, no important voices in the United States called for a withdrawal from the treaty, a testament to the attitude shift it had sparked.

While leaders on both sides were intent on reaching an agreement, the potential snares and pitfalls were not inconsiderable, especially since many parties on each side of the table would have welcomed a breakdown of the negotiations. Success required focus from the top, and here Kennedy played his role consummately. Harriman was given considerable authority in conducting the negotiations on the ground, combined with Kennedy's oversight and daily consultation. Harriman sent detailed cables to the State Department every day for consideration by a small White House team, including the president, Rusk, McNamara, and Sorensen. As Glenn Seaborg related, "Harriman, though a veteran of many troubleshooting missions for several presidents, was not prepared at first for the intensity of President Kennedy's interest in the detail of the negotiations."[33] Two and a half years in office had taught Kennedy the importance of controlling flows of information. Only a few people were shown Harriman's daily progress reports, and no copies were made or circulated.[34] Kennedy also kept in touch with key senators, thereby building the political support that would be needed for Senate ratification.

On July 25, the treaty was completed.[35] The three negotiating parties—the United States, Soviet Union, and the United Kingdom—initialed the treaty that day, with the formal signing ceremony following on August 5. The document is very short, not even three pages. The preamble states two purposes: the "speediest possible achievement of an agreement on general and complete disarmament," and "the discontinuance of test explosions of nuclear weapons for all time." Article I calls on the parties to prohibit all nuclear tests in the atmosphere, outer space, and underwater. Article II allows for amendments. Article III opens the treaty to all states for signature, and declares that the treaty enters into force after its ratification by the three original parties. Article IV declares the option for withdrawal.

When he received the happy news of the treaty's conclusion at his Hyannis Port compound in Cape Cod, Kennedy was already looking ahead to the next steps with the Soviets. William Foster, Director of the U.S. Arms Control and Disarmament Agency, recalled: "[We] had a relaxed drink or two in the balmy air of a summer's day. He stated that he felt that this was the most important thing he had accomplished thus far in his Administration—the achievement of a limitation on testing. He did appear elated. It was a result of his personal persistence . . . He would not give up before one more try because the stakes were so high."[36] Adrian Fisher, deputy director of the Arms Control and Disarmament Agency and Harriman's top aide in the Moscow negotiations, recounted: "I think he felt this was just a start in terms of a relationship with the Soviets . . . [H]e felt, 'Now you've got this one, let's get cracking on something else . . .' Generally speaking, his approach then was, 'Let's not rest now. This is a first step. Let's get cracking on something else.'"[37]

But first the treaty would face the U.S. public and the Senate in its ultimate test.

Chapter 7.

CONFIRMING THE TREATY

JOHN F. KENNEDY WAS aware that negotiating and sign-
ing the treaty was just half the battle. The ratification by two-thirds
of the U.S. Senate was the other half. The Senate defended its con-
stitutional prerogatives like a hawk. On many occasions the Senate had dashed the best hopes of a president after a major treaty
was signed. Kennedy would again draw on his vast powers of oratory and persuasion to seal the deal.

Woodrow Wilson's failure with the League of Nations offered
lessons for every subsequent president, and the lessons were especially relevant for Kennedy. He hailed from the same party as
Wilson, and was an heir of Wilsonian idealism, the belief in international institutions and treaties as the basis for international
peace. Kennedy was above all intent on avoiding Wilson's mistakes in dealing with the public and the Senate. He was determined from the start of the peace campaign to ensure that
everything agreed upon in negotiations would also be confirmed
in the Senate.

Norman Cousins later recalled what Kennedy had said about

Wilson when he met with a small group of opinion leaders to strategize about the campaign to win Senate ratification:

> Ever since Woodrow Wilson, he said, a President had to be cautious about bringing a treaty to the Senate unless he had a fairly good idea where the votes would come from. To get two-thirds of the Senate behind any issue was a difficult and dubious undertaking; to get it on a controversial treaty was almost in the nature of a miracle. He said that he could name fifteen senators who would probably vote against anything linked to Kennedy's name—"and not all of them are Republicans."[1]

Speaking to the Nation

The campaign for peace was based on winning the public's support for the treaty, and thereby pushing the Senate into agreement as well. Kennedy therefore began the campaign with a speech to the country and kept the public informed through press conferences, coverage of the Europe trip, and another major televised address on July 26, just one day after the treaty was initialed in Moscow. Beyond that, Kennedy met with national opinion leaders and actively supported the formation of the Citizens' Committee for a Nuclear Test Ban.[2] After serving as a go-between of Kennedy and Khrushchev in the spring, Norman Cousins helped to coordinate the campaign for public support in the summer. Cousins related a meeting with Kennedy:

> He reiterated the need for important business support and suggested a dozen names. He said that scientists such as James R. Killian [of MIT] and George Kistiakowsky [of Harvard] would be especially effective if

they could be recruited. He felt that religious figures, farmers, educators, and labor leaders all had key roles to play and mentioned a half dozen or more names in each category. Then he went down the list of states in which he felt extra effort was required.[3]

Kennedy played a very active role in strategizing and marshaling support from a wide cross-section of public figures. Most important, Kennedy spoke publicly, spoke compellingly, and spoke often. He used the presidential bully pulpit exquisitely.

With the treaty agreed upon in Moscow on July 25, Kennedy turned to the nation in a televised address.[4] Kennedy had mastered the medium, and used it to place himself directly in the living rooms and kitchens of families across the country. His remarks still radiate a power of intimacy and persuasion half a century later. He looked into the lens of the camera and the eyes of his countrymen.

"Good evening, my fellow citizens," he began. "I speak to you tonight in a spirit of hope. Eighteen years ago the advent of nuclear weapons changed the course of the world as well as the war. Since that time, all mankind has been struggling to escape from the darkening prospect of mass destruction on earth." Kennedy reminded his fellow citizens of the "vicious circle of conflicting ideology and interest," of how "[e]ach increase of tension has produced an increase of arms; each increase of arms has produced an increase of tension." He recalled how seemingly endless rounds of meetings on disarmament had "produced only darkness, discord, or disillusion."

"Yesterday," he said dramatically, "a shaft of light cut into the darkness."

Like a new beginning, first there was light. "Negotiations were concluded," announced Kennedy, "on a treaty to ban all nuclear tests in the atmosphere, in outer space, and under water." For the very first time since the dawn of the nuclear age, an agreement had been reached on bringing "the forces of nuclear destruction under international control."

As Kennedy described the treaty, he once again used the rhetorical tactic of beginning with a series of negatives—what the treaty was not. He would sell the treaty by underselling it. He would make it compelling as a first step on the journey to peace, not as an end point. He would not let the skeptics accuse him of starry-eyed diplomacy or, worse, appeasement. He would instead soberly inform his fellow Americans why it was right and prudent for them, for all Americans, to take this first step.

He also began with a clarification. Many plans, he noted, have been blocked by "those opposed to international inspection," obliquely referring to the Soviet Union. Onsite inspections are needed, he explained, only for underground tests. "The treaty initialed yesterday, therefore, is a limited treaty which permits continued underground testing and prohibits only those tests that we ourselves can police."

"We should also understand," said Kennedy, "that it has other limits as well." Signatories could withdraw from the treaty. "Nor does this treaty mean an end to the threat of nuclear war. It will not reduce nuclear stockpiles . . . it will not restrict their use in time of war." In fact:

> This treaty is not the millennium. It will not resolve all conflicts, or cause the Communists to forego their ambitions, or eliminate the dangers of war. It will not reduce our need for arms or allies or programs of assistance to others.

So what was it if it was *not* all of these things? "[I]t is an important first step—a step towards peace—a step towards reason—a step away from war."

Quietly, soberly, Kennedy answered the basic question that would be asked by any citizen: "[W]hat this step can mean to you and to your children and your neighbors." His answer had four parts:

"First, this treaty can be a step towards reduced world tension and broader areas of agreement." Other issues—general disarmament, a comprehensive test ban—were ultimate hopes, but not attained in this treaty.

"Second, this treaty can be a step towards freeing the world from the fears and dangers of radioactive fallout." Kennedy was careful not to oversell this point, though fallout was a cause of great public alarm. "The number of children and grandchildren with cancer in their bones, with leukemia in their blood, or with poison in their lungs might seem statistically small to some," but "the loss of even one human life . . . should be of concern to us all.

"Third, this treaty can be a step towards preventing the spread of nuclear weapons to nations not now possessing them." Kennedy spoke of the possibility that "many other nations" would soon have nuclear capacity, and he reminded his listeners that "if only one thermonuclear bomb were to be dropped on any American, Russian, or any other city . . . that one bomb could release more destructive power on the inhabitants of that one helpless city than all the bombs dropped in the Second World War":

> Neither the United States nor the Soviet Union nor the United Kingdom nor France can look forward to that day with equanimity. We have a great obligation, all four nuclear powers have a great obligation, to use whatever time remains to prevent the spread of nuclear weapons, to persuade other countries not to test, transfer, acquire, possess, or produce such weapons.
>
> This treaty can be the opening wedge in that campaign.

Kennedy then turned to the last point. "Fourth and finally, this treaty can limit the nuclear arms race in ways which, on balance,

will strengthen our Nation's security far more than the continuation of unrestricted testing." A nation's security, Kennedy reminded the country, "does not always increase as its arms increase,
when its adversary is doing the same." Once again, Kennedy explained the fundamental illogic of an arms spiral and the mutual
benefits of slowing it, something that he hoped the treaty would
help to do. Violations of the treaty—secret testing—would be possible, but the strategic gains would be slight and the costs to the
violator's reputation would be very high. In sum, the treaty, "in
our most careful judgment, is safer by far for the United States
than an unlimited nuclear arms race."

Throughout the address, Kennedy did not rely on a single line
of jargon, nor did he suggest that the public should simply adopt
the views of experts. All was laid out and explained simply and
precisely, down to the very choices that the negotiators had made
in the previous weeks. And so it was consistent and natural that
Kennedy would conclude by calling on all citizens to participate in
the upcoming Senate debate:

> The Constitution wisely requires the advice and consent
> of the Senate to all treaties, and that consultation has al-
> ready begun. All this is as it should be. A document which
> may mark an historic and constructive opportunity for
> the world deserves an historic and constructive debate.
>
> It is my hope that all of you will take part in that de-
> bate, for this treaty is for all of us. It is particularly for
> our children and our grandchildren, and they have no
> lobby here in Washington. This debate will involve mili-
> tary, scientific, and political experts, but it must be not
> left to them alone. The right and the responsibility are
> yours.

Kennedy concluded his address with the central theme: that we
must pursue a path of peace, one that is uncertain, risky, and chal-

lenging, but critical nevertheless. "No one can be certain what the future will bring . . . But history and our own conscience will judge us harsher if we do not now make every effort to test our hopes by action, and this is the place to begin. According to the ancient Chinese proverb, 'A journey of a thousand miles must begin with a single step.'"

And then came Kennedy's inimitable call to his countrymen:

> My fellow Americans, let us take that first step. Let us, if we can, step back from the shadows of war and seek out the way of peace. And if that journey is a thousand miles, or even more, let history record that we, in this land, at this time, took the first step.

Kennedy's was a call to action no less direct and stirring than Gandhi's call to Indians to step toward the sea to collect salt, and thereby free themselves from colonial rule. Kennedy used the word "step" fourteen times in the speech. He knew that peace would require a long journey, beyond a thousand miles and beyond a thousand days, but he cogently laid out the urgency of the first step: ratifying the Partial Nuclear Test Ban Treaty.

Avoiding Wilson's Blunders

Kennedy had Senate ratification in mind at every step of the negotiating process. He would not fall into Wilson's trap, succeeding in negotiating the treaty internationally but then failing to achieve its ratification domestically. Kennedy had an enormous advantage over Wilson, in addition to the obvious one of being able to learn from Wilson's mistakes. Kennedy, unlike Wilson, was a former senator. There was no way that Kennedy would forget the Senate's mores, and especially the determined way that senators would defend their prerogative to advise and consent on a major treaty.

From the start, Kennedy considered the domestic politics of the treaty from every vantage point, and pressed his advantage in every manner available. He knew that all parts of the treaty would face scrutiny and that he had to be ready for it. He knew that he would need solid bipartisan support to get two-thirds of the Senate, since some southern Democrats would vote against the treaty. He knew that powerful American thought leaders, if not handled with care, could derail the treaty. In short, he knew that ratification would require a precise campaign whose complexities would rival those of negotiating the treaty itself.

Kennedy had carefully pondered his main points of vulnerability. First, of course, were the European allies. If they bolted, the Senate would surely not go along. Thus, the trip to Europe was vital to bolster his political standing in Europe overall and specifically to solidify West German support for Kennedy's leadership of the Western alliance, and thus indirectly for the test ban treaty. Second was the military top brass. If the Joint Chiefs of Staff were hostile to the treaty, it would almost certainly fail in the Senate. Third were the natural Senate foes: hardline Republicans and southern Democrats. It was crucial for Kennedy to avoid making the treaty a partisan cause, since the northern and midwestern Democrats simply did not have the votes on their own. Fourth, top nuclear scientists could influence the Senate if they were hostile to a test ban. And fifth, of course, was the general public itself. Broad public support was not sufficient to ensure ratification, but it was surely necessary. Senators would have a much harder time rejecting the treaty in a political environment of broad public support.

Kennedy considered in detail each step of the treaty process from the point of view of ultimate Senate ratification. For example, should Kennedy, Khrushchev, and Macmillan sign the treaty at a summit meeting? Macmillan and Khrushchev wanted that,

Kennedy with the Joint Chiefs of Staff: Marine Corps general David M. Shoup, Army general Earle G. Wheeler, Air Force general Curtis E. LeMay, President Kennedy, chairman of the Joint Chiefs of Staff General Maxwell D. Taylor, and chief of naval operations Admiral George W. Anderson Jr. (January 15, 1963).

but Kennedy wisely demurred lest the treaty appear overly identified with Kennedy himself—something that had helped doom Wilson's League of Nations. He suggested to the other leaders that the foreign ministers sign the treaty, and in the case of the United States, that Secretary of State Rusk be backed by a bipartisan delegation of senators. After twisting the arms of some reluctant Republican senators who did not want to get too far out in front of the ratification process, Kennedy put together the bipartisan Senate delegation, and the treaty was duly signed in Moscow on August 5. The bipartisan delegation was a signal to the overall Senate, which would now take up consideration of the treaty.

The next issue was the support, or at least acquiescence, of the Joint Chiefs of Staff. "I regard the Chiefs as key to this thing," Kennedy told Senate majority leader Mike Mansfield. "If we don't get the Chiefs just right, we can . . . get blown." The chiefs have "always been our problem."[5] And they were indeed dragging their feet. Even the moderate chairman of the Joint Chiefs of Staff, General Maxwell Taylor, was unenthusiastic, but his reservations paled in comparison with those of the head of the air force, General Curtis LeMay.*

LeMay had tangled repeatedly and aggressively with Kennedy, especially during the Cuban Missile Crisis, when LeMay had wanted to bomb the Soviet missile sites. Roswell Gilpatric described that every time Kennedy "had to see LeMay, he ended up in a fit. I mean he just would be frantic at the end of a session with LeMay because, you know, LeMay couldn't listen or wouldn't take in, and he would make what Kennedy considered . . . outrageous proposals that bore no relation to the state of affairs in the 1960s."[6] LeMay had denigrated Kennedy's blockade strategy in the missile crisis as "appeasement." Now LeMay might try to derail the test ban treaty, too.

To forestall that danger, Kennedy accepted various reservations to the treaty suggested by the Joint Chiefs of Staff, albeit ones that would not require a renegotiation of terms. This was something that the uncompromising Wilson had refused to do when trying to persuade the Senate to approve his treaty. Kennedy met individually with each service chief and then with the Joint Chiefs as a group, and also had Secretary of State Dean Rusk, CIA director John McCone, and Atomic Energy Commission chairman Glenn Seaborg explain to the Joint Chiefs why each of them supported the treaty.[7] At a meeting on July 23, Kennedy agreed to four "safeguards" proposed by the Joint Chiefs:[8]

1. an "aggressive" underground test program
2. maintenance of "modern" weapons laboratories
3. readiness to resume atmospheric tests promptly
4. improvement of the U.S. capability to monitor
   Sino-Soviet nuclear activity

---

* The cigar-smoking LeMay served as the model for air force general Jack D. Ripper in the 1963 movie *Dr. Strangelove or: How I Learned to Stop Worrying and Love the Bomb.* Dallek, *An Unfinished Life,* 345.

The chiefs also consulted with the State Department on the political implications of the treaty. Two days after the treaty was signed, General Taylor wrote to Secretary of State Rusk to ask for his input:

The Joint Chiefs of Staff feel the need of your counsel and that of the Department of State in order to reach a thorough understanding of the implications and consequences of the implementation of this treaty. We recognize that the military considerations falling within our primary field of competence are not the exclusive determinants of the merits of a test ban treaty. In addition, important weight must be given to less tangible factors such as the effect upon world tensions and international relations.[9]

In the end, the Joint Chiefs were brought on board. Though they were somewhat reluctant converts, they abided by the guidance of the president and the foreign policy leaders on the geopolitical significance of the treaty, and felt satisfied with the safeguard provisions that Kennedy had agreed to.

The Senate Deliberates

All eyes turned toward the Senate on August 8, as President Kennedy submitted the treaty for the Senate's advice and consent. Two Senate committees held extensive hearings. The Senate Foreign Relations Committee, chaired by William Fulbright, began on August 12, and called an extensive roster of witnesses, including administration officials, nuclear scientists, and other public leaders. As recounted by Seaborg:

The committee hearings and the subsequent discussion on the Senate floor constituted one of the most intellectually demanding debates in the country's history. The matters considered included complex military, scientific, political, philosophical, and psychological questions.[10]

When the Joint Chiefs testified to the Senate regarding the treaty, each averred that he supported the treaty as long as the agreed-upon safeguards were in place. None of them broke ranks, though LeMay did tell Senator Goldwater that had the treaty not already been initialed, he might have opposed it even with the safeguards.[11] The chair of the Joint Chiefs, General Taylor, made it explicit that the Joint Chiefs had been part of the negotiating process throughout, and had had ample opportunity to contribute to the process. In their Senate testimony, the Joint Chiefs summarized in this way:

> Having weighed all of these factors, it is the judgment of the Joint Chiefs of Staff that, if adequate safeguards are established, the risks inherent in this treaty can be accepted in order to seek the important gains which may be achieved through a stabilization of international relations and a move toward a peaceful environment in which to seek resolution of our differences.[12]

Administration officials prepared their testimony in great detail. The White House also followed the testimony of high-profile hostile witnesses such as the atomic scientist Edward Teller, known as the "father of the hydrogen bomb," and encouraged scientists friendly to the treaty to respond to Teller's objections point by point.[13] The head of the Los Alamos Scientific Laboratory, Norris Bradbury, was particularly forceful, calling the treaty "the first sign of hope that international nuclear understanding is possible."[14] "If now is not the time to take this chance," he asked, "what combination of circumstances will ever produce a better time?"

President Eisenhower also came out publicly in support of the treaty, writing:

> Humankind is urgently seeking for some release from the tensions created by the global struggle between two opposing ideologies . . . Because each side possesses weapons of incalculable destructive power and with extraordinary efficiency in the means of delivery, world fears and tensions are intensified and, in addition, there is placed upon too much of mankind the costly burdens of an all-out arms race, both nuclear and conventional. As a consequence, any kind of agreement or treaty that plausibly purports to achieve even a modicum of relief from these burdens and tensions is eagerly welcomed by the peoples of the world.[15]

In the end, the Senate Foreign Relations Committee came down forcefully in favor of the treaty, by a vote of 16 to 1. A second set of hearings, by a subcommittee of the Senate Armed Services Committee, was antagonistic, voting 6 to 1 against the treaty, but with much less political consequence. From the hearings, the treaty moved to the Senate floor on September 9 for debate.

Once again, Kennedy did what Wilson had not: he paid full respect to the Senate leaders of both parties. Sorensen described Kennedy's personal engagement with senators on the issue:

> It paralleled few other efforts which he took place [sic] during the three years of the Kennedy presidency. He personally talked with a great many senators; he worked through the Senate leadership, the Vice President [Lyndon B. Johnson], the legislative liaison officers in the White House under Mr. [Lawrence F.] O'Brien and in the State Department under Mr. [Frederick G.] Dutton. He kept in daily touch with the tactics being used both

by the proponents and opponents of the treaty . . . He met, as I said, with the representatives of the coalition of organizations (which he had helped bring together, I believe) in support of the test ban treaty, counseled them as to how to spend their time and money, which senators they should see, what kind of effort should be made in Washington and back home, and so on. His own speeches and statements at the time were filled with references to this effort.[16]

Kennedy also sent a detailed letter on September 10 giving certain "unqualified and unequivocal assurance to the members of the Senate, to the entire Congress, and to the country."[17] In the letter, he reiterated his full backing for the four safeguards advocated by the Joint Chiefs of Staff, as well as for other considerations to address assorted concerns. These considerations included clear statements that the treaty did not affect the U.S. non-recognition of East Germany; that the treaty would not limit the use of nuclear weapons for the defense of the United States; and that the United States would seek necessary international approvals if needed for peaceful uses of nuclear explosions (such as for building canals).

In a masterstroke, he asked the Republican Senate minority leader, Everett Dirksen, to present the letter to the full Senate. In doing so, Dirksen also declared his own endorsement of the treaty, explaining that he would not like it written on his tombstone that "he knew what happened at Hiroshima but he did not take the first step."*[18]

The Senate floor debate occurred against the backdrop of a nation strongly mobilized in support of the treaty. Opinion surveys

---

* Yet there may have been a bit more to it as well. Rumor had it that Kennedy's assent to halt a corruption investigation of a former top Eisenhower official was another powerful lure to Dirksen for his support of the treaty. Seaborg, *Kennedy, Khrushchev, and the Test Ban,* 280.

Kennedy signs the Partial Nuclear Test Ban Treaty (October 7, 1963).

showed that the administration's round-the-clock efforts at building public support, including Kennedy's personal meetings with leading newspaper editors and major interest groups, had paid off handsomely. A Harris poll soon after Kennedy's July 26 speech gave 53 percent "unqualified approval" of the treaty, 29 percent "qualified approval," and 17 percent opposed.[19] By September, the unqualified approval rating had risen to 81 percent. A Gallup poll showed 63 percent approval, 17 percent opposed, and 20 percent without an opinion.

The final Senate vote came on September 24, with a resounding tally of 80 to 19 in favor of the treaty. Kennedy released a statement that day declaring the ratification "a welcome culmination of this effort to lead the world once again to the path of peace . . . I congratulate the Senate for its actions."[20] Through personal leadership, Kennedy had achieved his historic success, with an enormous and bipartisan margin.

Global Approval

The treaty was opened for signature by other countries immediately upon the signing by the Soviet Union, United States, and the United Kingdom on August 5. As of that date, more than thirty other countries had already signaled their intention to accede. By the time the treaty went to the Senate floor on September 9, ninety countries had signed. The White House closely tracked international opinion.[21] The United States had hoped that the Soviet Union could sway China to sign, and the Soviet Union hoped the United States would do the same for France. Neither came to pass. China and France were determined that they would not be deterred from pursuit of their own nuclear arsenal.

Kennedy Reaches Out to World Leaders

The hopes of the world were lifted by the signing of the treaty. At this moment of high hopes, on September 20, Kennedy stepped up to the rostrum of the UN General Assembly to address fellow world leaders and senior statesmen, two long years after he had last spoken there.[22]

"We meet again in the quest for peace," he began. "Twenty-four months ago, when I last had the honor of addressing this body, the shadow of fear lay darkly across the world . . . Those were anxious days for mankind." But this time, Kennedy brought good news:

> Today the clouds have lifted a little so that new rays of hope can break through . . . And, for the first time in 17 years of effort, a specific step has been taken to limit the nuclear arms race.
>
> I refer, of course, to the treaty to ban nuclear tests in the atmosphere, outer space, and under water—

Kennedy delivers his address to the UN General Assembly (September 20, 1963).

concluded by the Soviet Union, the United Kingdom, and the United States—and already signed by nearly 100 countries. It has been hailed by people the world over who are thankful to be free from the fears of nuclear fallout.

Once again, he underscored the positive by beginning with the negative. "The world has not escaped from the darkness. The long shadows of conflict and crisis envelop us still." "But," he continued, "we meet today in an atmosphere of rising hope, and at a moment of comparative calm. My presence here today is not a sign of crisis, but of confidence."

This was Kennedy's chance to speak heart to heart with his global counterparts. "I am not here to report on a new threat to the peace or new signs of war. I have come to salute the United Nations and to show the support of the American people for your daily deliberations." He repeated the theme from the American University address that peace is a process, not some grand declaration or magic formula. "Peace," he emphasized, "is a daily, a weekly, a monthly process, gradually changing opinions, slowly eroding old barriers, quietly building new structures. And however undramatic the pursuit of peace, that pursuit must go on."

As he had told the American people eight weeks earlier, he told the General Assembly that the treaty was merely the first step of a journey, not the end:

> Today we may have reached a pause in the cold war— but that is not a lasting peace. A test ban treaty is a milestone—but it is not the millennium. We have not been released from our obligations—we have been given an opportunity. And if we fail to make the most of this moment and this momentum—if we convert our new-found hopes and understandings into new walls and weapons of hostility—if this pause in the cold war merely leads to its renewal and not to its end—then the indictment of posterity will rightly point its finger at us all. But if we can stretch this pause into a period of cooperation—if both sides can now gain new confidence and experience in concrete collaborations for peace—if we can now be as bold and farsighted in the control of deadly weapons as we have been in their creation—then surely this first small step can be the start of a long and fruitful journey.

Kennedy emphasized the universal responsibility for peace. As he had told the Irish parliamentarians, it is for small nations as well as large ones:

> The task of building the peace lies with the leaders of every nation, large and small. For the great powers have no monopoly on conflict or ambition. The cold war is not the only expression of tension in this world—and the nuclear race is not the only arms race. Even little wars are dangerous in a nuclear world. The long labor of

peace is an undertaking for every nation—and in this effort none of us can remain unaligned. To this goal none can be uncommitted.

And as he had two years earlier from the same rostrum, Kennedy described what he hoped would be the next steps of the journey:

> I believe, therefore, that the Soviet Union and the United States, together with their allies, can achieve further agreements—agreements which spring from our mutual interest in avoiding mutual destruction.
>
> There can be no doubt about the agenda of further steps. We must continue to seek agreements on measures which prevent war by accident or miscalculation. We must continue to seek agreements on safeguards against surprise attack, including observation posts at key points. We must continue to seek agreement on further measures to curb the nuclear arms race, by controlling the transfer of nuclear weapons, converting fissionable materials to peaceful purposes, and banning underground testing, with adequate inspection and enforcement. We must continue to seek agreement on a freer flow of information and people from East to West and West to East.

Yet Kennedy raised the world's sights still higher. Peace could make possible a new surge of global problem solving, a new attention to the world's common interests:

> The effort to improve the conditions of man, however, is not a task for the few. It is the task of all nations—acting alone, acting in groups, acting in the United Nations, for plague and pestilence, and plunder and pollution,

the hazards of nature, and the hunger of children are the foes of every nation. The earth, the sea, and the air are the concern of every nation. And science, technology, and education can be the ally of every nation.

Never before has man had such capacity to control his own environment, to end thirst and hunger, to conquer poverty and disease, to banish illiteracy and massive human misery. We have the power to make this the best generation of mankind in the history of the world—or to make it the last.

Remarkably, Kennedy set out a technological agenda that would presage the world's sustainable development efforts for decades to come:

- A world center for health communications under the World Health Organization could warn of epidemics and the adverse effects of certain drugs as well as transmit the results of new experiments and new discoveries.
- Regional research centers could advance our common medical knowledge and train new scientists and doctors for new nations.
- A global system of satellites could provide communication and weather information for all corners of the earth.
- A worldwide program of conservation could protect the forest and wild game preserves now in danger of extinction for all time, improve the marine harvest of food from our oceans, and prevent the contamination of air and water by industrial as well as nuclear pollution.
- And, finally, a worldwide program of farm productivity and food distribution, similar to our country's "Food

for Peace" program, could now give every child the food he needs.

Kennedy also drew the deep links between peace and human rights, and the challenge of ending long-standing discrimination in the United States and elsewhere. Building on the American University speech of June 10 and his civil rights address of June 11, he said:

> I know that some of you have experienced discrimination in this country. But I ask you to believe me when I tell you that this is not the wish of most Americans—that we share your regret and resentment—and that we intend to end such practices for all time to come, not only for our visitors, but for our own citizens as well.
>
> I hope that not only our Nation but all other multiracial societies will meet these standards of fairness and justice. We are opposed to apartheid and all forms of human oppression.

Toward the end of his remarks, Kennedy reminded his fellow leaders that peace "does not rest in charters and covenants alone. It lies in the hearts and minds of all people":

> So let us not rest all our hopes on parchment and on paper; let us strive to build peace, a desire for peace, a willingness to work for peace, in the hearts and minds of all our people. I believe that we can. I believe the problems of human destiny are not beyond the reach of human beings.

"Two years ago," Kennedy recalled, "I told this body that the United States had proposed, and was willing to sign, a limited test ban treaty. Today that treaty has been signed. It will not put an end to war. It will not remove basic conflicts. It will not secure freedom for all." But, he said, "it can be a lever."

> [A]nd Archimedes, in explaining the principles of the lever, was said to have declared to his friends: "Give me a place where I can stand—and I shall move the world."
>
> My fellow inhabitants of this planet: Let us take our stand here in this Assembly of nations. And let us see if we, in our own time, can move the world to a just and lasting peace.

Kennedy had moved the world. In the twelve months since October 1962 he had kept his gaze on peace. He had held fast to his faith in humanity. He had trusted the virtue of America's adversaries. And he had been vindicated: mankind was not doomed, gripped by forces beyond its control.

Against the Odds

After the fact, historical events have an air of inevitability, because we forget the many contingencies—personal leadership, luck, timing, and accidents—that made them possible. Yet we must remember: there was nothing inevitable about achieving a test ban treaty. Here is how the historian Vojtech Mastny described the policy scene as of early 1963:

> Even though Kennedy wanted a rapprochement between the superpowers, he, like Khrushchev, could not easily afford it. The European allies, while favoring détente in principle, were nervous about a superpower deal over their heads. Within the United States, members of Congress pilloried the Soviet Union for its behavior in Cuba and cited it as evidence that Khrushchev

could not be trusted. The Joint Chiefs of Staff were dead set against a nuclear test ban, arguing that it would compromise the U.S. strategic deterrent. Senator Everett Dirksen branded the talks on nuclear testing an "exercise not in negotiation . . . but in give-away." Within NATO, West German Chancellor Konrad Adenauer denounced the test ban as an invitation to Soviet blackmail. Among Washington's major allies, only the British favored the ban, urging U.S. concessions to make it possible.[23]

The Partial Test Ban Treaty was signed and ratified for one overwhelming reason: Kennedy campaigned for it. He was a gifted campaigner: in six campaigns between 1946 and 1960, he had not lost a single one.* And he triumphed again in the summer of 1963.

When the treaty was adopted, both Kennedy and Khrushchev believed that further easing of tensions would soon follow. This was to be, as Kennedy said so frequently, the first step on a journey. In a short period of time, there would be others: cultural exchanges; the famous "hotline" for direct communication and crisis management, agreed upon ten days after the Peace Speech; the sale of $250 million worth of wheat to the Soviet Union;[24] potential cooperation in space, as Kennedy had offered in his September 1963 address to the UN General Assembly.

But the world's hopes were followed by despair just eight weeks later when Kennedy was assassinated. The outpouring of grief was beyond any easy reckoning. For all of the limitations of his brief time in office, for all of his missteps and incomplete work, Kennedy had touched the hearts of people around the world, had caused them to share a goal of peace and to move irresistibly toward it.

* The closest he ever came to losing an election was at the 1956 Democratic National Convention, where he was nominated for vice president but finished second in the balloting to Senator Estes Kefauver. At the same convention, Kennedy's future negotiator of the test ban treaty, W. Averell Harriman, finished second in the presidential balloting to nominee Adlai Stevenson, Kennedy's ambassador to the UN.

In his condolence letter to President Johnson, Khrushchev wrote that Kennedy's death was a grievous loss for the United States, and "that the gravity of this loss is felt by the whole world, including ourselves, the Soviet people . . . it was an awareness of the great responsibility for the destinies of the world that guided the actions of the two Governments—both of the Soviet Union and of the United States—in recent years. These actions were founded on a desire to prevent a disaster and to resolve disputed issues through agreement with due regard for the most important, the most fundamental interests of ensuring peace."[25] Jacqueline Kennedy wrote to Khrushchev:

> I know how much my husband cared about peace, and how the relation between you and him was central to this care in his mind. He used to quote your words in some of his speeches—"In the next war the survivors will envy the dead." You and he were adversaries, but you were allied in a determination that the world should not be blown up. You respected each other and could deal with each other.[26]

Not long after Kennedy's death came Khrushchev's political demise. Kennedy's bond with Khrushchev had certainly helped to sustain the Russian leader in 1963 even after Khrushchev's missteps in Cuba and elsewhere. With Kennedy gone, Khrushchev's hold on power soon weakened. Within four months, Leonid Brezhnev began to plot Khrushchev's ouster, which occurred in October 1964, eleven months after Kennedy's death. The peaceful ouster itself reflected a forward movement of Soviet politics, as Khrushchev himself acknowledged to a confidant:

> I'm old and tired. Let them cope by themselves. I've done the main thing. Could anyone have dreamed of telling Stalin that he didn't suit us anymore and suggest-

ing he retire? Not even a wet spot would have remained where we had been standing. Now everything is different. The fear is gone, and we can talk as equals. That's my contribution. I won't put up a fight.[27]

With the protagonists gone so quickly from the scene, Kennedy's peace initiative was prematurely put to the test. Would the Partial Test Ban Treaty prove to be the first step of a journey toward peace, even one of a thousand miles? Or would it prove to be a fleeting moment, a brief Camelot in the *Sturm und Drang* of the raging Cold War?

Chapter 8.

THE HISTORIC MEANING
OF KENNEDY'S PEACE
INITIATIVE

AS KENNEDY HIMSELF had acknowledged, the treaty was not the millennium: it did not end conflicts, lead to peace, or halt the arms race. In many areas, the spirit of cooperation that began in 1963 was overwhelmed by the ongoing and intrinsic dynamics of the Cold War. Yet it made a lasting difference, one that inspires and challenges us in our own time.

Doomsday

In 1947, the *Bulletin of the Atomic Scientists* created the Doomsday Clock, which indicated how close humanity was to global disaster, or "midnight." In 1947 the clock was placed at seven minutes to midnight; by 1949, humanity had given up four minutes of its margin, with the clock at just three minutes to midnight. The Soviet Union had become a nuclear power: the nuclear arms race was on. By 1953, humanity had given up another minute: the clock stood at just two minutes to midnight. Both sides had thermonuclear weapons. "Only a few more swings of the pendulum, and, from Moscow to Chicago, atomic explosions will strike midnight for Western civilization."[1]

Eisenhower had left the clock at seven minutes till midnight, the same position as in 1947, the start of the Cold War. Kennedy's peace initiative pushed the minute hand back to twelve minutes before midnight in 1963, a new margin of safety. The treaty contributed to a decade of détente. Yet matters began to unravel once again after 1972, and the minute hand moved forward perilously over the coming dozen years, reaching just three minutes to midnight in 1984. Then, at a moment of high peril, the momentum shifted again toward peace with the accession to power of Mikhail Gorbachev. Gorbachev's reforms brought the Cold War to a peaceful end in 1991, pushing the minute hand back to seventeen minutes before midnight, the largest buffer of safety since the start of the nuclear era. Yet even that dramatic gain has proved to be evanescent. In our time, the minute hand has rushed forward once again to just five minutes to midnight.

Figure 1. Doomsday Clock: Minutes to Midnight, 1947–2012.

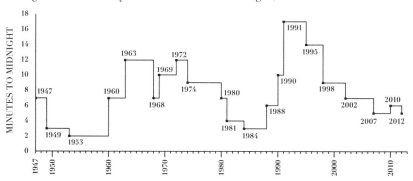

Throughout these ups and downs, the Partial Test Ban Treaty helped to keep humanity away from the precipice. Its most *direct* long-lasting impact was on nuclear proliferation, just as Kennedy had hoped. Yet its larger significance was as conclusive proof that cooperation between the superpowers was possible, a fundamental lesson and legacy of enduring significance.

The Partial Success of Non-Proliferation

The Partial Test Ban Treaty (PTBT) proved that agreements could be reached and honored by both sides, and in this way it gave a powerful impetus to a series of arms control treaties along the lines that Kennedy had outlined in his 1961 speech at the UN. Kennedy had suggested six steps: (1) a test ban; (2) a stop to the production and transfer to other nations of fissionable material; (3) a stop to the transfer of control over nuclear weapons to states that do not own them; (4) a prohibition of nuclear weapons in space; (5) a gradual destruction of existing weapons; and finally (6) a halt to the production of strategic nuclear delivery vehicles and their gradual destruction. Steps 1 through 4 were substantially achieved before the end of the Cold War, and the progress that was made owes much to the Partial Test Ban Treaty, while steps 5 and 6 have proved to be far more elusive until recently.

Most important, the PTBT made possible the signing of the Non-Proliferation Treaty (NPT) in 1968, a treaty of profound significance for global security.[2] For both Kennedy and Khrushchev, the specter of a massive proliferation of nuclear weapons was the main driving force behind the test ban treaty. Their hope was that a ban on nuclear tests would slow or stop the ability of other countries to become nuclear powers. Kennedy famously worried aloud that fifteen to twenty countries would become nuclear powers by the 1970s:

The reason why we keep moving and working on this question, taking up a good deal of energy and effort, is because personally I am haunted by the feeling that by 1970, unless we are successful [with implementing a test ban treaty], there may be 10 nuclear powers instead of 4, and by 1975, 15 or 20 . . . I see the possibility in the 1970s of the President of the United States having to face a world in which 15 or 20 or 25 nations may have these weapons. I regard that as the greatest possible danger and hazard.[3]

Cynics from the 1950s onward argued that a non-proliferation treaty would be a scrap of paper at best, and an opportunity for subterfuge at worst. Why would a mere paper promise to abjure nuclear weapons take precedence over what a government saw as its national interest? Yet in retrospect, the treaty has been relatively effective, far more so than the doubters imagined.

At the time of signing, there were five nuclear powers: the United States, Soviet Union, the United Kingdom, France, and China—the five permanent members of the UN Security Council. From then till now, dozens more have had the technological know-how to become nuclear powers. Yet since the signing of the NPT, not a single signatory has become a nuclear state, and only four non-signatories have joined the nuclear club: Israel, India, Pakistan, and North Korea. One country, South Africa, embarked on a nuclear program but later renounced it, signing the Non-Proliferation Treaty in 1991. Several others, including South Korea and Taiwan, were successfully pressured by the United States to drop nuclear programs.

By 1995, the NPT's great value was evident, and the initial twenty-five-year period of the treaty was extended indefinitely. How exactly had it accomplished its ends? According to one leading observer, Thomas Graham, it was the new "international norm against nuclear-weapon proliferation established by the NPT":

In 1960, after the first French nuclear-weapon test, there were banner newspaper headlines, "Vive La France." Yet, by the time of the first Indian nuclear explosion in 1974, the test was done surreptitiously, India received worldwide condemnation and New Delhi hastened to explain that this had been a "peaceful" test. What had intervened was the NPT. It converted the acquisition of nuclear weapons by a state from an act of national pride in 1960 to an act contrary to international law in 1974.[4]

Nuclear proliferation remains a grave threat, of course. The Indian and Pakistani armies face off against each other in Kashmir, and the two countries have often bitter relations. North Korea continues to threaten South Korea, and the threats include unleashing its missiles. Israel lives defiantly among unfriendly neighbors and views its nuclear weapons as a critical deterrent. Iran is widely suspected by many of building a nuclear bomb and has been declared by the UN to be in violation of its NPT obligations regarding inspection of nuclear sites.

Moreover, the basic commitment of the nuclear powers to move toward a nuclear-free world has not been fulfilled. The nuclear powers retain vast arsenals, and though they are markedly reducing their stockpiles of weapons, they have not convinced the world that they are truly moving to a nuclear-free world. This lack of adequate follow-through by the major nuclear powers seriously undermines the norm of non-proliferation, and could eventually convince still more countries to become nuclear powers. Fortunately, some of America's leading foreign policy voices have recently endorsed the realism and importance of a nuclear-free world in line with the original aspirations of the NPT.[5]

The Era of Détente

On a political level, the test ban treaty also marked a watershed. The U.S.-Soviet showdown in Europe was significantly calmed for the next twenty years, until tensions soared again in the early 1980s with President Ronald Reagan's arms buildup and the U.S.

deployment of new intermediate-range missiles in Europe.[6] Most important and positive, Germany would never again play the same role as the flashpoint of superpower controversy.

Three major elements produced the thaw in U.S.-Soviet relations. First, the clear decision by Kennedy to keep nuclear weapons away from Germany resolved one of the greatest sources of Soviet fear and U.S.-Soviet controversy. Kennedy had faced down Adenauer, and Johnson finally pulled the plug once and for all in 1964 on the MLF proposal for nuclear sharing among NATO members. Second, the final break in Soviet-Chinese relations in the 1960s, which deteriorated to the point of armed conflict along the border in 1969, dramatically changed the geopolitical context of the Cold War. The U.S.-Soviet relationship became one side of a three-sided relationship that now included China as a nuclear power independent of, and indeed hostile to, the Soviet Union. And third, the Kennedy-Khrushchev achievement of the PTBT created a powerful worldwide demonstration that peaceful coexistence was possible and could contribute to fruitful diplomatic achievements with mutual gains.

Sadly, the easing of external tensions did not lead quickly to an easing of the repressive Soviet system, either internally or in Eastern Europe. Internally, the Soviet Union entered a long period of economic stagnation in the 1960s and then outright decline in the 1980s. The continuing arms race, which shifted resources from the civilian economy to the arms buildup, certainly poisoned morale inside the Soviet Union. And when the reduction of East-West tensions led Czechoslovakia to experiment with its Prague Spring in 1968, Soviet tanks crushed the attempted liberalization. Throughout Eastern Europe, any signs of liberalization were met by a harsh crackdown.

The high point of U.S.-Soviet détente came in the early 1970s. President Richard Nixon and Chairman Leonid Brezhnev negotiated agreements on strategic arms control as well as the basic principles to govern bilateral relations. Simultaneously, Nixon began the process of normalizing relations with China with his remarkable visit to Beijing in 1972. The United States recognized the People's Republic of China in 1979, which in turn was followed by a surge of bilateral economic relations.

The Strategic Arms Limitation Talks (SALT I) from 1969 to 1972 culminated in three main agreements signed by Nixon and Brezhnev. The first, the Anti-Ballistic Missile (ABM) Treaty, put a limit on the deployment of ABM systems. The second, the Interim Agreement on the Limitation of Strategic Offensive Arms, froze the number of strategic ballistic missiles for a five-year period. The third, the Basic Principles of Relations Between the United States and the USSR, established key principles to govern bilateral relations.[7]

The SALT I talks were immediately followed by the SALT II talks, which extended from 1972 to 1979.[8] SALT II aimed to limit the number of nuclear delivery systems (missiles and bombers), the construction of new missile launchers, and the deployment of new types of strategic offensive weapons systems. In short, it attempted a more general limit on the still-burgeoning quantity and quality of strategic nuclear weapons. Yet although the SALT I treaty was overwhelmingly ratified by the U.S. Senate, SALT II was never approved; the Soviet invasion of Afghanistan six months after it was signed led President Jimmy Carter to withdraw the treaty from Senate consideration. More generally, the gains of détente had already been deeply undermined even before the Soviet invasion of Afghanistan.

Why? Most important was the drift of political power on both sides back toward the hardliners, showing again how the dynamics of confrontation can easily get out of the control of leaders on both sides, as hardliners on each side gradually get the upper hand, playing on and stoking the fears of the public. American hardliners claimed that the continuing Soviet nuclear buildup after 1963 was proof of Soviet intentions of a first-strike capacity and global domination. They blasted Presidents Nixon, Ford, and

Carter for endangering U.S. security by making deals with the enemy. Hardliners on the Soviet side claimed the same about the United States. Each side viewed the other side's actions as aggressive rather than defensive.

Ronald Reagan came to office in 1981 vowing to avenge supposed U.S. losses in the Cold War and to restore U.S. power and influence. He quickly ushered in major increases in U.S. military outlays, stepped up CIA operations on many fronts, created a new missile defense initiative and a proposal to upgrade intermediate nuclear forces in Europe, and engaged in aggressive rhetoric against the "evil" Soviet empire. The period from 1981 to 1985 marked a clear intensification of the Cold War.

Reagan's plans to build new missile defenses (the Strategic Defense Initiative, or "Star Wars" as it came to be called), and to deploy upgraded intermediate-range nuclear missiles in Europe, in response to preceding Soviet deployments of its SS-20 missiles, were particular causes of tension. For two or three years, with Europe bitterly divided over whether the United States should in fact deploy new intermediate-range missiles, and as Reagan kept up a blistering rhetorical attack on the Soviet Union, many Soviet officials surmised that the United States was moving toward war. The mood briefly seemed to approach the worst days of confrontation during the late 1950s and early 1960s.

Many on the Soviet side were strongly convinced that Reagan was preparing for nuclear first strike. The Doomsday Clock had it right when it put the minute hand at just three minutes to midnight in 1984. As the clock keepers noted at the time:

> U.S.-Soviet relations reach their iciest point in decades. Dialogue between the two superpowers virtually stops. "Every channel of communications has been constricted or shut down; every form of contact has been attenuated or cut off. And arms control negotiations have been reduced to a species of propaganda," a concerned Bul-

letin informs readers. The United States seems to flout the few arms control agreements in place by seeking an expansive, space-based anti-ballistic missile capability, raising worries that a new arms race will begin.[9]

The Nuclear Arms Race

Kennedy's goals 5 and 6, which called for the actual dismantling of nuclear weapons systems, did not come to fruition during the Cold War, though there has been notable progress since 1991. Some argue that the test ban treaty might even have accelerated the nuclear arms buildup in the first years after signing. Before 1963, atmospheric testing was contentious and politically difficult. Once the tests went underground, the public ignored them. The nuclear weapons industries in the United States and the Soviet Union might have thereby gained rather than lost room to ma-

Figure 2. U.S. and Soviet Nuclear Warheads, 1946–2012.[10]

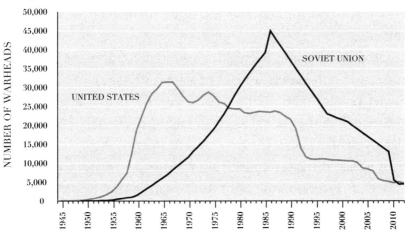

Notes: For the years 1946–2009, total nuclear warheads. For 2010–2012, total warheads minus those awaiting dismantlement. Note that all data are estimates, including uncertainties about the numbers of warheads awaiting dismantlement.

neuver. The arms control agreements reached in the 1970s and 1980s had some restraining effects on the size and type of the nuclear arsenals, notably on anti-ballistic missile systems, but the number of warheads on the U.S. side remained extraordinarily high, around 25,000–30,000 until the mid-1980s. The Soviet Union even surpassed the United States, its stockpile rising from 4,000 warheads in 1963 to 45,000 in 1986.[11] This dramatic Soviet buildup from 1963 to 1986 is shown in Figure 2.

On the U.S. side, Kennedy promised the Joint Chiefs of Staff an immediate step-up in underground testing as part of the "safeguard" provisions. As soon as the treaty was initialed, the United States commenced Operation Niblick, a series of forty-one underground tests at a Nevada testing site. From then until 1992, when the program was finally shut down, the United States conducted a total of 684 underground tests. These tests were used to design and promote new weapons systems and for the continued (and expensive) upgrading of the U.S. nuclear arsenal.[12] While the total number of U.S. nuclear weapons declined slightly from around 28,000 in 1963 to around 21,000 in 1990, the delivery capability of the U.S. arsenal soared with new deployments of increasingly sophisticated land-based and submarine-based ballistic missiles.

On the Soviet side, the nuclear arms buildup after 1963 was dramatic and unrelenting until Gorbachev's accession to power. The Soviet political and military leadership felt deeply aggrieved by the success of the United States in forcing a reversal of the missile deployment in Cuba. Their conclusion was that the U.S. advantage in nuclear warheads and delivery systems had been crucial to America's success in the crisis itself. Many Soviet leaders pledged that the Soviet Union would never again face that kind of humiliating retreat as the result of its nuclear inferiority. Between 1963 and 1986, the Soviet Union closed the numerical gap in warheads (as of 1978), and then soared beyond the U.S. numbers, with the total number of Soviet ICBMs rising by a factor of around 14, warheads by a factor of around 10, and total megatons by a factor of around 6.[13]

This mutual arms buildup was avoidable, but very strong political leadership would have been required on both sides to face down the vested interests of each country's military-industrial complex. Robert Jervis's "security dilemma" was clearly and repeatedly at play. Though each side characterized its own buildup as defensive—intended only to keep up with the other side—the other side invariably interpreted each new arms deployment as an attempt at nuclear dominance, perhaps even at a nuclear first strike. Again and again, each side used "worst-case" analyses to justify the next upward ratchet of the arms race.

In the end, the arms race up till 1991 deeply undermined the long-term interests of both sides. The military outlays were enormous, and came at the neglect of domestic needs. The Soviet Union erred catastrophically by crushing its domestic economy in order to make room for its massive nuclear buildup. Over two decades, from the mid-1960s to the mid-1980s, the Soviet military-industrial complex in effect suffocated the civilian economy, drained civilian morale and economic incentives, and thereby contributed to the economic implosion that destroyed the Soviet Union itself.

Realists on both sides tried to break the arms race, but hardliners and the military-industrial complexes on each side easily outmaneuvered them. National Security Adviser and Secretary of State Henry Kissinger, for example, tried to slow the arms buildup in the mid-1970s through arms control agreements. In a 1974 press conference in Moscow he stated in exasperation:

> One of the questions which we have to ask ourselves as a country is what in the name of God is strategic superiority? What is the significance of it, politically, militarily, operationally, at these levels of numbers? What do you do with it?[14]

Some claim that the collapse of the Soviet Union was the ultimate vindication of the U.S. armaments buildup during the Cold War. Did not the United States, after all, end up bankrupting the Soviet state? This may be the case, but there is also good reason to believe that the Soviet Union might have reformed itself in a peaceful yet far less tumultuous manner, as have China and almost all other state-run economies. Or the Soviet Union might have collapsed under the weight of economic failure even without the nuclear arms race. A less tumultuous collapse would have come at much lower cost to the people of the Soviet Union, clearly, but also at lower cost and risk for the rest of the world as well. Moreover, the dramatic collapse of the Soviet state and economy in the late 1980s and early 1990s, and the resulting chaos, lawlessness, and corruption, have surely detracted from the human and security gains the world might otherwise have enjoyed from the demise of the Soviet system.

Much better news on arms control came at the very end of the Cold War, when the United States and Soviet Union, in its final days, began a process to reduce sharply the number of nuclear warheads and delivery systems. START I, the Strategic Arms Reduction Treaty signed in 1991, limited the total warheads on each side to around 6,000, with a ceiling of around 1,600 missiles and bombers.[15] In 1993, START II limited the deployment of MIRVs (multiple independently targeted warheads on a single missile).[16] The most significant step has come recently, with the New START treaty (2010), which caps the number of warheads at around 1,550 (with technicalities affecting the count) and strategic delivery systems at 700.[17] The sharp decline in the number of warheads from the mid-1980s to 2012 can be seen in Figure 2.

These notable accomplishments are in line, finally, with Kennedy's aspirations of a half century earlier. The numbers are coming down to around a twentieth or less of their peak numbers at the height of the Cold War arsenals. The declining numbers have given some realistic hope that progress can be made toward full nuclear disarmament, a goal that, as I mentioned earlier, is now articulated by some of America's most experienced foreign policy figures.* Of course the prospects for a nuclear-free world are made far more complicated by the continuing threats of proliferation to new countries such as Iran and North Korea, and even to rogue groups. Nuclear arms control, alas, can no longer be achieved by just a handful of countries.

Proxy Wars

An even more deadly disappointment was the lack of positive spillover from relaxed U.S.-Soviet tensions into the peaceful resolution of regional conflicts. If anything, the United States and Soviet Union continued to spur and participate directly in proxy wars around the world as part of their ongoing Cold War confrontation and rivalry. The United States started this post-PTBT pattern with Johnson's dramatic escalation of the Vietnam War in 1964–1965. The result was a devastating ten-year ordeal in which the United States committed more than 500,000 troops, killed more than one million and perhaps up to two million Vietnamese

---

* Most notable in the United States has been the leadership of former secretary of state Henry Kissinger, former U.S. senator Sam Nunn, former secretary of state George Shultz, and former secretary of defense William Perry, in urging a nuclear-free world. They have published a series of highly noted and influential op-ed pieces since 2007 calling for practical steps to reduce the nuclear threat and ultimately achieve a world without nuclear weapons. In their first piece they stated: "Reassertion of the vision of a world free of nuclear weapons and practical measures toward achieving that goal would be, and would be perceived as, a bold initiative consistent with America's moral heritage. The effort could have a profoundly positive impact on the security of future generations. Without the bold vision, the actions will not be perceived as fair or urgent. Without the actions, the vision will not be perceived as realistic or possible. We endorse setting the goal of a world free of nuclear weapons and working energetically on the actions required to achieve that goal." George P. Shultz, William J. Perry, Henry A. Kissinger, and Sam Nunn, "A World Free of Nuclear Weapons," Wall Street Journal, January 4, 2007.

civilians, lost more than 50,000 U.S. soldiers in combat, and bled itself militarily, financially, socially, and emotionally for decades thereafter.[18]

Alongside the surging nuclear arsenals, proxy wars became the new emblems of the Cold War. These wars—in South America, the Middle East, and elsewhere—were often fueled and supported by secret CIA operations on the U.S. side, and corresponding acts of secret destabilization and militarization on the Soviet side. As such, they constituted wars within wars, often effectively unsupervised by political authorities and feeding the most extreme ideological, militaristic, or even personal agendas on each side.

Vietnam was America's greatest blunder of the Cold War; Afghanistan proved to be even more devastating for the Soviet Union. In the mid- to late 1970s, impoverished Afghanistan fell into paroxysms of instability in a series of coups and counter-coups, exacerbated by regional instability, including the Iranian Revolution, the Indian nuclear tests in 1974, the 1979 Egypt-Israel peace agreement (which was viewed by the Soviet Union as a threat to Soviet influence in the Middle East), and extreme tensions with neighboring Pakistan. In 1979, the Soviet Union made the fateful decision to intervene in order to support a Soviet-favored faction. This in turn provoked the United States to support an international force of Islamic fighters, the mujahideen, who were organized and funded by the CIA.[19] The Afghanistan war proved even more of a drain on the finances, the morale, and the military capacity of the Soviet Union than the Vietnam War was for the United States. The bloody and costly debacle, ended by Mikhail Gorbachev only in 1989, was one of the factors leading to the Soviet economic crisis of the late 1980s, which in turn helped to precipitate the collapse of the Soviet Union in 1991.

The End of the Cold War

The end of the Cold War was shocking on several fronts. Most amazing, of course, was its rapidity, predicted by almost nobody.[20] The Berlin Wall collapsed in 1989, and the Soviet Union disappeared in 1991, a dozen years after the Soviet invasion of Afghanistan, which had been seen by some in the United States as a high-water mark of Soviet global power. A country with more than 40,000 nuclear weapons and a massive internal security apparatus collapsed in disarray.

There are countless views about the causes of the end of the Soviet Union, and these conflicting interpretations may never be resolved. To the American anti-communist hardliners, Ronald Reagan's aggressive posture was the decisive margin of victory. According to this view, he had re-instilled an arms race that the Soviet Union could not match. Later, he had charmed Gorbachev, so the story goes, into concessions that otherwise would not have been made.

Yet the demise of the Soviet Union was vastly more complex than this. Many other leaders played important roles. John Paul II, the Polish pope, was another decisive force, re-instilling hope, nationalism, anti-communist fervor, and faith in a country reeling from martial law under the Soviet shadow. Great and brave dissidents from Alexander Solzhenitsyn and Andrei Sakharov to Lech Walesa, Vaclav Havel, and others in Eastern Europe had a profound effect on undermining the legitimacy of the Soviet regime. The 1975 Helsinki Declaration, including Article VII on human rights, created a far more powerful norm of human rights across the communist world than the Soviet leaders (and nearly everybody else) imagined possible at the time. Here was Churchill and Kennan's ultimate insight at play: that even a tightly closed Soviet system would ultimately wither under exposure to a more open world. So too did unexpected events accelerate the collapse of the

Soviet state, notably the 1986 Chernobyl nuclear disaster, and the collapse of world oil prices in the 1980s.

In the end, the Soviet Union and its economic and political system unraveled through a concatenation of economic and geopolitical events. Gorbachev deserves the world's supreme credit as the highest practitioner of peace, the greatest statesman of the age. As the Soviet economy spiraled downward, as Soviet power in Eastern Europe came unstuck, as the disastrous Afghan War ate away at the nation's morale, as internal dissension in Russia and throughout the non-Russian states threatened the sovereignty of the nation itself, Gorbachev himself held firm: No violence would be deployed to hold the Soviet Union in place. This was the fundamental departure from all that had gone before. For this, Gorbachev merits the world's highest approbation. One wonders how many politicians of any ilk in any part of the world would have displayed such fortitude for peace in the midst of such earth-shaking events.

Yet the fairest verdict of all is that the Soviet Union collapsed under its own weight, not because of the Cold War, not because of a single personality or particular adverse events, but because the organization of the Soviet economy and political system proved to be fatally flawed. The planned economy was simply a bad idea, one that had shown certain early merits—namely in rapid heavy industrialization—but was beset by far deeper long-term flaws: the lack of incentives, the lack of technological dynamism, the inability to plan an entire economy, and the relative isolation of the Soviet economy from technological advances occurring in the rest of the world.

It is useful to recall two great thinkers as we today ponder the demise of the Soviet Union. The first is John Maynard Keynes, the greatest political economist of the twentieth century. He had occasion to visit Russia in 1925, to form a judgment about the new communist order just taking hold. His economic judgment was damning, complete, and fully vindicated in time:

On the economic side I cannot perceive that Russian Communism had made any contribution to our economic problems of intellectual interest or scientific value. I do not think that it contains, or is likely to contain, any piece of useful economic technique which we could not apply, if we chose, with equal or greater success in a society which retained all the marks, I will not say of nineteenth-century individualistic capitalism, but of British bourgeois ideals.[21]

Still, Keynes wondered whether beneath the "cruelty and stupidity of New Russia some speck of the ideal may lay hid." Could the revolution usher in new ideas regarding human values that would prove of lasting worth in social organization?

A second great thinker, George Kennan, believed that the answer was no. The grand architect of America's containment policy, the author of the influential Long Telegram and "The Sources of Soviet Conduct,"* believed that containment of the Soviet Union would succeed not because America's military power would vanquish the communist foe but because the flaws of the Soviet system would cause the system to change from within. Kennan, like Keynes, believed that the Soviet model was deeply unsound, and that internal discontents would eventually bring the system down. The U.S. challenge was to meet Soviet threats along the way, so that in the course of time the messianism of the Soviet system would be forced back to reality:

It would be an exaggeration to say that American behavior unassisted and alone could exercise a power of life and death over the Communist movement and bring about the early fall of Soviet power in Russia. But

---

* These two documents, the first a diplomatic cable in 1946 and the second an anonymously authored piece for *Foreign Affairs* magazine in 1947, were hugely influential in setting the new U.S. policy framework of containment.

the United States has it in its power to increase enormously the strains under which Soviet policy must operate, to force upon the Kremlin a far greater degree of
moderation and circumspection than it has had to observe in recent years, and in this way to promote ten-
dencies which must eventually find their outlet in either
the breakup or the gradual mellowing of Soviet power.
*For no mystical, Messianic movement—and particularly
not that of the Kremlin—can face frustration indefinitely
without eventually adjusting itself in one way or another
to the logic of that state of affairs.*[22]

Kennan surmised "that the possibility remains (and in the opinion of this writer it is a strong one) that Soviet power, like the
capitalist world of its conception, bears within it the seeds of its
own decay, and that the sprouting of these seeds is well advanced."
The great challenge of U.S.-Soviet relations, wrote Kennan, was
therefore not of a military variety, still less of a nuclear arms race.
It was one of values:

The issue of Soviet-American relations is in essence a
test of the overall worth of the United States as a nation
among nations. To avoid destruction the United States
need only measure up to its own best traditions and
prove itself worthy of preservation as a great nation.[23]

A Summing Up

The ultimate meaning of Kennedy's peace initiative therefore lay
not in the tactics, nor in the success of the test ban treaty, nor even
in the laying of the groundwork for nuclear non-proliferation and
further arms agreements. Nor are the undeniable and even dev-

astating setbacks—the Vietnam War, the dangerous and costly nuclear arms race, the terrible and costly proxy wars, the military-industrial complex, and the dangers of a security state—proofs that Kennedy's peace initiative failed. For the legacy of the peace initiative must be seen where Kennan put them squarely: as a test of values.

Kennedy put forward three definitions of peace. Peace, he said, is the necessary rational end of rational men; peace is a process, a way of solving problems; and peace is a human right, a right to live in safety, free from fear. Kennedy claimed that our common humanity could transcend our obvious differences, and that the world could be made safe for diversity. And he urged steadfastness, to take peace a step at a time, and to realize that our work on earth "will not . . . be finished in the first 1,000 days, nor in the life of this Administration, nor even perhaps in our lifetime on this planet. But let us begin."

Time proved to be a great elixir. By staying alive, by avoiding the abyss, the world moved beyond the Cold War and beyond the closed Soviet system. As Kennedy predicted in his Peace Speech and his address to the Irish Dáil, the enmities that nearly destroyed the world in 1962 seem ancient and nearly incomprehensible to us today. The extreme ideologues, fanatics, and messianists of past political faiths, whether communist or anti-communist, were all proved wrong, though vastly too many lives were lost in the process. Those who believed in the common fate of humanity were vindicated, and we owe our survival to them.

Chapter 9.

LET US TAKE OUR STAND

THE MOST IMPORTANT LESSON that we learn from John Kennedy is to fashion the future out of our rational hopes, not our fears. He was the first to deny the baseless hopes of idle dreamers:

> I am not referring to the absolute, infinite concept of universal peace and good will of which some fantasies and fanatics dream. I do not deny the value of hopes and dreams but we merely invite discouragement and incredulity by making that our only and immediate goal.

But he was also for seeing things as they might be, and asking, "Why not?"

What other lessons do we learn from his leadership that we can apply in our own time? We are inspired by Kennedy's repeated urging that each generation must take up the great challenges of its time. Kennedy relished the causes of his day, especially the defense of liberty: "I do not believe," he declared, "that any of us would exchange places with any other people or any other generation." Through all of his speeches, the challenge is always specifically drawn. "Let us take our stand here . . . in our own time." When we behold Kennedy's energy and confidence in problem solving, we are stirred to bring the same daring commitment to problem solving in our own time.

Kennedy inspired others because he demanded the best from them. He did not tell Americans, Russians, or Berliners what to do; he told them, instead, what they could achieve. He created his enduring legacy not by making easy promises but by calling for difficult sacrifices. Fifty years on, we are still moved by his case for going to the moon:

> We choose to go to the moon in this decade and do the other things, not because they are easy, but because they are hard, because that goal will serve to organize and measure the best of our energies and skills, because that challenge is one that we are willing to accept, one we are unwilling to postpone, and one which we intend to win, and the others, too.[1]

Kennedy insisted repeatedly that our fates are intertwined, and that our own well-being depends on an indissoluble chain of human well-being. "Freedom is indivisible," he told the Berliners, "and when one man is enslaved, all are not free." He called for moral reflection on the basic equality of our fates: "The heart of the question is whether all Americans are to be afforded equal rights and equal opportunities, whether we are going to treat our fellow Americans as we want to be treated." And he bid us to be courageous in accepting moral responsibility:

> A great change is at hand, and our task, our obligation, is to make that revolution, that change, peaceful and

constructive for all. Those who do nothing are inviting shame as well as violence. Those who act boldly are recognizing right as well as reality.[2]

For Kennedy, the common fate of humanity was not only a moral vision of universal rights but a practical key to statecraft and political leadership. Because humanity shares a common fate and common aspirations, it is possible to step into the shoes of one's adversary and to understand the problem from the adversary's point of view. Dehumanizing an opponent was not only a moral error but a tactical one, too. As Kennedy said in the Peace Speech, the adversary is a fellow human being, with the same interests, inhabiting the same small planet, breathing the same air, caring for their own children in the same way that we do. And, as he stated, no political or social system is so repugnant that its people should be regarded as lacking in virtue.

Kennedy knew that vision was not enough, and that a general call to peace and well-being would accomplish little. Kennedy spoke "in this time and place" about specific challenges, whether they be peace, race relations, the race to the moon, or other causes. As a politician and statesman, he looked relentlessly for a practical path, a next step toward the goal. He gave us the best single piece of management advice that I know, one that I admire so much that I'll quote it again:

> By defining our goal more clearly, by making it seem more manageable and less remote, we can help all people to see it, to draw hope from it, and to move irresistibly towards it.

Clear goals are vital for many reasons: they create shared objectives, help to specify the means, and unite the public in action. Defining goals is indeed the most important single job of leadership, since without them there can be nothing but cacophony. Only once a goal has been made clear can the practical work of problem solving begin. And with a clear goal, a leader can galvanize society to bold actions, whether going to the moon or making peace with the Soviet Union. To win the public's backing for the moon mission, Kennedy quoted William Bradford, a founder of the Plymouth Bay Colony, to the effect that "all great and honorable actions are accompanied with great difficulties, and both must be enterprised and overcome with answerable courage." Kennedy's quest for peace was the same: a clear goal, courageously embraced; and practical steps to achieve it, with milestones on nuclear testing, non-proliferation, arms control, and so forth.

Kennedy recognized another obstacle. In opening the Peace Speech, he called peace "the most important topic on earth." Yet he noted that "the pursuit of peace is not as dramatic as the pursuit of war, and frequently the words of the pursuers fall on deaf ears." Here is a dismaying truth: the most important topic on earth may fall on deaf ears! We are hardwired for drama, for competition, for the struggle to survive. Even when we cooperate, we often do it for the benefit of our own group—so our group can be stronger than the others. Kennedy himself took advantage of this inclination: when he called space the ultimate frontier, an adventure for all humanity, he motivated Americans in part by declaring that America would be first in space, thus appealing to our competitive nature.

Global cooperation is more elusive than cooperation within clans, families, tribes, and nations. How do we mobilize attention to and efforts at cooperation on a global scale, when the challenge is not "us" versus "them"? Kennedy made progress in this direction, by emphasizing our common humanity and the mutual benefits of cooperation. We can use his example, his ideas, and his oratory as we struggle to achieve global cooperation in our time.

Our Generation's Challenge

Each generation is born into a new world, and faces new challenges. Those born at the start of the twentieth century grappled
with the Great Depression and two devastating world wars. When
the torch was passed to Kennedy's generation, one that had been
"tempered by war and disciplined by a hard and bitter peace," as
he noted in his inaugural, the challenge was to face the Cold War
and the new realities of nuclear weapons.

Our generation's challenge is different. Fundamentally, it is
the challenge of globalization, of a crowded, interconnected world
of seven billion–plus people (more than twice the three billion
alive at Kennedy's inauguration). Our world society is now intertwined more tightly than ever before—by markets, by technology,
by social networks—and at the same time faces unprecedented
stresses on a global scale: environmental change, resource scarcity,
mass migration, vast disparities of income and wealth, and destabilizing technological change. These realities of globalization are
daunting and unprecedented. They are not clearly perceived or
understood by much of humanity. Our political and economic
systems have not yet begun to cope with these new realities.

Between 1991 and 2012, as we see in Figure 1, the *Bulletin of the
Atomic Scientists* moved the Doomsday Clock from seventeen
minutes till midnight—its safest level since the atomic age began
in 1945—to just five minutes from midnight. We are once again at
risk of spiraling out of control, as we were for a dozen years after
1947 and for a dozen years more after 1972. Once again we will
need a peace initiative, one suitable for our own time.

The dire message of the Doomsday Clock in 2012 is no longer
about the confrontation of superpowers. Rather, it warns us about
the complexities of globalization and sustainable development
and the proliferation of nuclear weapons:

The challenges to rid the world of nuclear weapons, harness nuclear power, and meet the nearly inexorable climate disruptions from global warming are complex and interconnected. In the face of such complex problems, it is difficult to see where the capacity lies to address these challenges. Political processes seem wholly inadequate; the potential for nuclear weapons use in regional conflicts in the Middle East, Northeast Asia, and South Asia are alarming; safer nuclear reactor designs need to be developed and built, and more stringent oversight, training, and attention are needed to prevent future disasters; the pace of technological solutions to address climate change may not be adequate to meet the hardships that large-scale disruption of the climate portends.[3]

Sustainable Development
as Our Common Goal

———

We define our generation's goal more clearly by placing it in the context of sustainable development. The dominant political priority in nearly every society in the world today is economic advancement. Every government in the world puts the strength of the economy at the top of its agenda. Its political survival, and its ability to compete with adversaries, depends on economic success. Yet sustainable development as a doctrine contends that the pursuit of economic gain alone cannot suffice for human well-being and security. There should be three objectives, not one. The economy should advance. The economic gains should be broad-based or "inclusive," cutting across different parts of society, different ethnic and minority groups, different classes, and benefiting women as well as men. And the gains should be sustainable in terms of resource use and the conservation of ecosystems. In short, we should aim not for economic growth alone, but for inclusive and environmentally sustainable growth. To accomplish this, governments also require a modus vivendi with the powerful corporate sector, which needs to play by the rules of sustainable development (for example, desisting from pollution and deforestation) without trying to use its lobbying muscle to write the rules for its narrow financial benefit.

The challenge of sustainable development all too often falls on deaf ears, as Kennedy warned about the challenge of peace. Sustainable development surely lacks the drama of the global war on terror. But like Kennedy's peace initiative, sustainable development can actually save lives in vast numbers and promote global prosperity, something that wars do not do. And as with arms control, the public's interest in sustainable development has been limited and episodic. Just as arms control negotiations went through countless fruitless sessions in New York, Geneva, and elsewhere before yielding results, sustainable development has been our own generation's putative commitment for more than twenty years. It was introduced to the world's consciousness by the Brundtland Commission of 1987, and introduced into global law at the Rio Earth Summit in 1992. Yet it has stuck mainly as a concept of theoretical and specialist interest, not as a broad-based and practical venture.

Sustainable development can alleviate global tensions and solve global problems if, following Kennedy's suggestions, the goals of sustainable development are defined more clearly and made more manageable and less remote. In our time we could at least take steps toward solutions to the great sustainable development challenges we face.

Peace Through Sustainable Development

We need to reconceive the challenge of peace itself. Making peace is a political and social process, but the ability to sustain peace depends on economic development. Impoverished countries fall into violence, conflict, and civil war with far greater frequency and predictability than do stable, prosperous societies. For this reason the United States launched the Marshall Plan in 1947, which provided economic assistance to postwar Europe, as a way to forestall Soviet political advances and consolidate democracies in Western Europe in the context of extreme economic duress. As Secretary of State George Marshall explained, "Our policy is directed not against any country or doctrine but against hunger, poverty, desperation and chaos. Its purpose should be the revival of a working economy in the world so as to permit the emergence of political and social conditions in which free institutions can exist."[4] U.S. leaders were fully aware that the Bolsheviks came to power in 1917 in the context of Russian economic chaos, and Hitler came to power in 1933 in the context of 25 percent German unemployment.

This most basic lesson has been lost on recent policymakers. Today's conflicts are found mostly in impoverished or economically destabilized countries. Consider Mali, Somalia, Yemen, and Afghanistan, four of the world's poorest countries. The United States is now engaged in covert or overt military action in all four. Many thousands of civilians die each year. And violent extremists are taking advantage of this instability to make bases for regional operations. Poverty opens the way for conflict, while conflict leads to a further downward spiral of impoverishment. The result is a poverty-violence trap, in which poverty and violence become chronic and mutually reinforcing.

Western intervention in these places has tended to be almost entirely military in character, though the problems they face can rarely be solved by military means, and never by military means alone. I know this bias firsthand, as I've tried for years to encourage some measure of development assistance from the United States, which would ultimately be cheaper than military spending. It has almost always been a losing cause. Poor places are treated as foreign policy irrelevancies until they succumb to violence and terror; then they are treated as military and security threats. There seems to be little Western policy in between these two extremes: complete neglect followed by panic and drones.

Looking at a map of the world's conflict zones, one is struck by the extent to which they are currently concentrated in the African and Asian drylands, an eight-thousand-mile swath that runs west to east from Senegal to Afghanistan, including the African Sahel, the Horn of Africa, the Arabian Peninsula, Western Asia, and Central Asia. The violence in this region is most commonly attributed to what Samuel Huntington famously called the "clash of civilizations," the alleged front line in an ongoing struggle between Islam and Christianity. Yet there are many peaceful Muslim places in the world, and many of them border Christian places. I suggest that Islamic extremism in the dryland belt is more symptom than cause. One of the deep causes is poverty against a backdrop of severe ecological stress, rising populations, diminishing rainfall, and a growing frequency of droughts and famines.

Recognizing that economic development is a key to peace and democratization, Kennedy launched the Peace Corps, the Alliance for Progress in Latin America, and similar development programs in other parts of the developing world. In his inaugural address, he addressed the people in the "huts and villages across the globe struggling to break the bonds of mass misery":

> We pledge our best efforts to help them help themselves, for whatever period is required—not because the Communists may be doing it, not because we seek their

votes, but because it is right. If a free society cannot help the many who are poor, it cannot save the few who are rich.

Yet foreign assistance was never popular with the American public or political leaders. The Marshall Plan itself barely passed Congress. And after Kennedy, foreign assistance faced a long-term decline in interest and effort. Once the Cold War was over, foreign aid plummeted even further, since the "strategic case" for aid had supposedly disappeared. The share of aid as a percentage of GDP, just 0.1 percent, was lower in 1998 than at any time since 1947. Following the terrorist attacks of September 11, 2001, aid rose slightly, especially to "allies" in the Middle East and Central Asia, but still remained low by historical and comparative standards, at about 0.2 percent of GDP.[5]

To reestablish aid as a pillar of foreign policy—thus saving money, lives, and future grief—would therefore require a change in the American mindset. Americans need to understand that impoverished foreign nations, even those that are ostensibly "foes," respond positively to practical aid—for health, education, and infrastructure—that is given openly and generously. In addition to providing practical and sometimes lifesaving help, aid signals human respect and a recognition of the commonality of human interests. Aid guided by the precepts of sustainable development would lead America back to true problem solving, the kind that a policy based on drone missiles can never accomplish. If we use our science, technology, and development experience to take on the challenge of basic economic development in the bereft places of the world—from Mali to Somalia, from Yemen to Afghanistan— our country itself would also be a huge beneficiary. We would win diplomatic allies, trading partners, and friends and supporters in the villages and cities of Africa and Asia, and for the right reasons.

Cooperative Problem Solving

Convincing an adversary or a competitor that we share aims and interests isn't easy. Trust is typically low, and there are ample reasons to bluff. Trust is even lower when countries have been adversaries for years or decades. It would have been much easier if Kennedy had needed to make peace with Canada, but he needed to do it with the Soviet Union, the state that had threatened America's very survival just months earlier.

We have learned many lessons from Kennedy's experience and its aftermath. We learned that only those leaders with a holistic and empathetic view are able to achieve success in complex negotiations with an adversary. Otherwise, the pessimists, hardliners, and fearmongers on each side can create self-fulfilling prophecies of failure. Kennedy therefore had to assert his leadership among his own colleagues just as much as with Khrushchev.

Another basic lesson is this: The path to success lies in the nature of the process of negotiation and mutual accommodation itself. Kennedy and Khrushchev signed agreements in 1963 because by then they knew and trusted each other, in part because of the bluster, bluffs, and near disasters that had come before. They had exchanged dozens of letters and suffered the consequences of many misunderstandings. By 1963 each had arrived at a realization he could not have had earlier: their situations were symmetrical. They each sought peace with the other despite a mood of militarism, the skepticism of the generals and hardliners, the vested interests of the military-industrial complex, and the interests and opportunism of their political competitors.

In the academic sphere, where many battles are also surprisingly bitter ("because," as the saying goes, "the stakes are so low"), the great economist (and Kennedy adviser) Paul Samuelson offered his own wisdom on the art of persuasion. He said that to convince another academic of a point, "give him a half-finished theorem." That is, let the other person reach his or her own conclusions, not through bluster, but through independent inquiry, guided by a half-finished product.

I want to urge a similar approach to the practical work of sustainable development. One of the reasons for the bitterness between Israelis and Palestinians, Indians and Pakistanis, Americans and Iranians, and other conflicting parties, is the almost complete lack of practical experience in solving problems together, working on "half-finished theorems." How easy it is to dehumanize one's adversaries when you peer at them through the lens of a drone, rather than work beside them in some common endeavor. And consider how many of our problems today are problems that cross national boundaries, and how easy it would be to share the burden and excitement of problem solving as well. Israelis and Palestinians share a small sliver of land facing increasing drought and depletion of freshwater resources. So far, Israel has faced this problem by commandeering a disproportionate share of the region's scarce water supply, but climate and demographic forecasts convince us that this is a losing battle for both sides. The dwindling freshwater resources will not sustain the combined populations of the two sides. Many (including me) have discussed this issue at length with Israelis and with Palestinians. Yet they have rarely discussed it with each other.

President Obama was on to something important in Cairo in 2009 when he proposed the establishment of a set of scientific centers of excellence "in Africa, the Middle East and Southeast Asia, and the appointment of new Science Envoys to collaborate on programs that develop new sources of energy, create green jobs, digitize records, clean water, and grow new crops."[6] This is the right approach. It echoes Kennedy's remarkable call for scientific collaboration in his speech to the UN General Assembly in 1963. Disappointingly, till now Obama's vision remains only that, a vision. It is high time to fulfill it, since surely it would mark a step toward peace.

And as always with trip wires of war, we may not have much time. The United States and Iran, for example, seem now to be on a relentless collision course, though the two countries could find much common ground if they tried. Iran is home to great culture, history, and know-how that could help to improve conditions not only in its own region, but in other parts of the world as well. Engagement, joint problem solving, and an honest negotiation over political differences would be vastly more fruitful and prudent than a military faceoff and the possibility of outright conflict.

Leadership and History

We owe our very lives to John Kennedy's grace under pressure in October 1962. We owe the eventual end of the Cold War in part to his ability to forge a measure of trust and respect between Americans and Russians in 1963, the final year of his life. Between then and now, though, we've squandered enormous opportunities. Millions have died needlessly in proxy wars with no real purpose; trillions of dollars, enough to end human poverty in all its forms, have instead been wasted on the Cold War arms races and outright conflicts.

Historians have long debated the great theme of whether people and societies can truly help to steer their fate. Are we but the flotsam on the turbulent seas of technological and social change, rising and sinking in waters beyond our control? Or, as Kennedy insisted, can man be as big as he wants? Is Kennedy right that no problem of human destiny is beyond human beings?

Not every moment of history is equally pregnant with the possibility of constructive choice. Some times are times of stasis that resist change. Others are periods of great flux, in which individual acts of leadership can make a profound difference for good or ill. Deep economic and geopolitical crises are such periods. At the height of the Cold War, with its potential for total destruction,

Kennedy had the opportunity to exercise choice, and he showed us how it could be done.

The stakes were so high in 1963 in large part because of the new technological realities, the new face of war in the nuclear age. As Kennedy noted in his inaugural address, man now held "in his mortal hands the power to abolish all forms of human poverty and all forms of human life." We have been struggling to save ourselves ever since, and that struggle continues until today.

At such a hinge of history, individuals can make a vast difference, and Kennedy was fully aware of the high stakes. His struggle was with the genie of nuclear power, and the unknowns of coexistence with a communist superpower. "With a good conscience our only sure reward, with history the final judge of our deeds, let us go forth to lead the land we love, asking His blessing and His help, but knowing that here on earth God's work must truly be our own."

Now it is our turn. We still confront the nuclear genie and the thousands of warheads that continue to threaten human survival. We are still challenged by the lack of trust within and between societies. We have developed and mastered remarkable new technologies but still flounder in the art of self-preservation. We still threaten ourselves with our own destruction, whether with our armaments or through the world's remarkable economic productivity coupled with a still-reckless disregard for the natural environment.

We know that our tasks are large, but so too are the acts of past leadership that inspire us and encourage us on our way. We have been granted the lessons of John Kennedy's peace initiative, and the gift of his and Ted Sorensen's words for our age and beyond. We are not gripped by forces beyond our control. We too can be as big as we want. We too can take our stand and move the world.

AMERICAN UNIVERSITY
COMMENCEMENT ADDRESS

June 10, 1963

President Anderson, members of the faculty, board of trustees, distinguished guests, my old colleague, Senator Bob Byrd, who has earned his degree through many years of attending night law school, while I am earning mine in the next 30 minutes, distinguished guests, ladies and gentlemen:

It is with great pride that I participate in this ceremony of the American University, sponsored by the Methodist Church, founded by Bishop John Fletcher Hurst, and first opened by President Woodrow Wilson in 1914. This is a young and growing university, but it has already fulfilled Bishop Hurst's enlightened hope for the study of history and public affairs in a city devoted to the making of history and to the conduct of the public's business. By sponsoring this institution of higher learning for all who wish to learn, whatever their color or their creed, the Methodists of this area and the Nation deserve the Nation's thanks, and I commend all those who are today graduating.

Professor Woodrow Wilson once said that every man sent out from a university should be a man of his nation as well as a man of his time, and I am confident that the men and women who carry the honor of graduating from this institution will continue to give from their lives, from their talents, a high measure of public service and public support. "There are few earthly things more beautiful than a university," wrote John Masefield in his tribute to English universities—and his words are equally true today. He did not refer to towers or to campuses. He admired the splendid beauty of a university, because it was, he said, "a place where those who hate ignorance may strive to know, where those who perceive truth may strive to make others see."

I have, therefore, chosen this time and place to discuss a topic on which ignorance too often abounds and the truth too rarely perceived. And that is the most important topic on earth: peace. What kind of peace do I mean and what kind of a peace do we seek? Not a Pax Americana enforced on the world by American weapons of war. Not the peace of the grave or the security of the slave. I am talking about genuine peace, the kind of peace that makes life on earth worth living, and the kind that enables men and nations to grow, and to hope, and build a better life for their children—not merely peace for Americans but peace for all men and women, not merely peace in our time but peace in all time.

I speak of peace because of the new face of war. Total war makes no sense in an age where great powers can maintain large and relatively invulnerable nuclear forces and refuse to surrender without resort to those forces. It makes no sense in an age where a single nuclear weapon contains almost ten times the explosive force delivered by all the allied air forces in the Second World War. It makes no sense in an age when the deadly poisons produced by a nuclear exchange would be carried by wind and water and soil and seed to the far corners of the globe and to generations yet unborn.

Today the expenditure of billions of dollars every year on

weapons acquired for the purpose of making sure we never need them is essential to the keeping of peace. But surely the acquisition of such idle stockpiles—which can only destroy and never create—is not the only, much less the most efficient, means of assuring peace. I speak of peace, therefore, as the necessary, rational end of rational men. I realize the pursuit of peace is not as dramatic as the pursuit of war, and frequently the words of the pursuers fall on deaf ears. But we have no more urgent task.

Some say that it is useless to speak of peace or world law or world disarmament, and that it will be useless until the leaders of the Soviet Union adopt a more enlightened attitude. I hope they do. I believe we can help them do it. But I also believe that we must reexamine our own attitudes, as individuals and as a Nation, for our attitude is as essential as theirs. And every graduate of this school, every thoughtful citizen who despairs of war and wishes to bring peace, should begin by looking inward, by examining his own attitude towards the possibilities of peace, towards the Soviet Union, towards the course of the cold war and towards freedom and peace here at home.

First examine our attitude towards peace itself. Too many of us think it is impossible. Too many think it is unreal. But that is a dangerous, defeatist belief. It leads to the conclusion that war is inevitable, that mankind is doomed, that we are gripped by forces we cannot control. We need not accept that view. Our problems are manmade; therefore, they can be solved by man. And man can be as big as he wants. No problem of human destiny is beyond human beings. Man's reason and spirit have often solved the seemingly unsolvable, and we believe they can do it again. I am not referring to the absolute, infinite concept of universal peace and good will of which some fantasies and fanatics dream. I do not deny the value of hopes and dreams but we merely invite discouragement and incredulity by making that our only and immediate goal.

Let us focus instead on a more practical, more attainable peace, based not on a sudden revolution in human nature but on a gradual evolution in human institutions—on a series of concrete actions and effective agreements which are in the interest of all concerned. There is no single, simple key to this peace; no grand or magic formula to be adopted by one or two powers. Genuine peace must be the product of many nations, the sum of many acts. It must be dynamic, not static, changing to meet the challenge of each new generation. For peace is a process—a way of solving problems.

With such a peace, there will still be quarrels and conflicting interests, as there are within families and nations. World peace, like community peace, does not require that each man love his neighbor, it requires only that they live together in mutual tolerance, submitting their disputes to a just and peaceful settlement. And history teaches us that enmities between nations, as between individuals, do not last forever. However fixed our likes and dislikes may seem, the tide of time and events will often bring surprising changes in the relations between nations and neighbors. So let us persevere. Peace need not be impracticable, and war need not be inevitable. By defining our goal more clearly, by making it seem more manageable and less remote, we can help all people to see it, to draw hope from it, and to move irresistibly towards it.

And second, let us reexamine our attitude towards the Soviet Union. It is discouraging to think that their leaders may actually believe what their propagandists write. It is discouraging to read a recent, authoritative Soviet text on military strategy and find, on page after page, wholly baseless and incredible claims, such as the allegation that American imperialist circles are preparing to unleash different types of war, that there is a very real threat of a preventive war being unleashed by American imperialists against the Soviet Union, and that the political aims—and I quote—"of the American imperialists are to enslave economically and politically the European and other capitalist countries and to achieve world domination by means of aggressive war."

Truly, as it was written long ago: "The wicked flee when no man pursueth."

Yet it is sad to read these Soviet statements, to realize the extent of the gulf between us. But it is also a warning, a warning to the American people not to fall into the same trap as the Soviets, not to see only a distorted and desperate view of the other side, not to see conflict as inevitable, accommodation as impossible, and communication as nothing more than an exchange of threats.

No government or social system is so evil that its people must be considered as lacking in virtue. As Americans, we find communism profoundly repugnant as a negation of personal freedom and dignity. But we can still hail the Russian people for their many achievements in science and space, in economic and industrial growth, in culture, in acts of courage.

Among the many traits the peoples of our two countries have in common, none is stronger than our mutual abhorrence of war. Almost unique among the major world powers, we have never been at war with each other. And no nation in the history of battle ever suffered more than the Soviet Union in the Second World War. At least 20 million lost their lives. Countless millions of homes and families were burned or sacked. A third of the nation's territory, including two thirds of its industrial base, was turned into a wasteland—a loss equivalent to the destruction of this country east of Chicago.

Today, should total war ever break out again—no matter how— our two countries will be the primary target. It is an ironic but accurate fact that the two strongest powers are the two in the most danger of devastation. All we have built, all we have worked for, would be destroyed in the first 24 hours. And even in the cold war, which brings burdens and dangers to so many countries, including this Nation's closest allies, our two countries bear the heaviest burdens. For we are both devoting massive sums of money to weapons that could be better devoted to combat ignorance, poverty, and disease. We are both caught up in a vicious and dangerous cycle, with suspicion on one side breeding suspicion on the other, and new weapons begetting counter-weapons. In short, both the United States and its allies, and the Soviet Union and its allies, have a mutually deep interest in a just and genuine peace and in halting the arms race. Agreements to this end are in the interests of the Soviet Union as well as ours. And even the most hostile nations can be relied upon to accept and keep those treaty obligations, and only those treaty obligations, which are in their own interest.

So let us not be blind to our differences, but let us also direct attention to our common interests and the means by which those differences can be resolved. And if we cannot end now our differences, at least we can help make the world safe for diversity. For in the final analysis, our most basic common link is that we all inhabit this small planet. We all breathe the same air. We all cherish our children's futures. And we are all mortal.

Third, let us reexamine our attitude towards the cold war, remembering we're not engaged in a debate, seeking to pile up debating points. We are not here distributing blame or pointing the finger of judgment. We must deal with the world as it is, and not as it might have been had the history of the last 18 years been different. We must, therefore, persevere in the search for peace in the hope that constructive changes within the Communist bloc might bring within reach solutions which now seem beyond us. We must conduct our affairs in such a way that it becomes in the Communists' interest to agree on a genuine peace. And above all, while defending our own vital interests, nuclear powers must avert those confrontations which bring an adversary to a choice of either a humiliating retreat or a nuclear war. To adopt that kind of course in the nuclear age would be evidence only of the bankruptcy of our policy—or of a collective death-wish for the world.

To secure these ends, America's weapons are nonprovocative, carefully controlled, designed to deter, and capable of selective use. Our military forces are committed to peace and disciplined in self-restraint. Our diplomats are instructed to avoid unnecessary irritants and purely rhetorical hostility. For we can seek a relaxation of tensions without relaxing our guard. And, for our part, we do not need to use threats to prove we are resolute. We do not need to jam foreign broadcasts out of fear our faith will be eroded. We are unwilling to impose our system on any unwilling people, but we are willing and able to engage in peaceful competition with any people on earth.

Meanwhile, we seek to strengthen the United Nations, to help solve its financial problems, to make it a more effective instrument for peace, to develop it into a genuine world security system— a system capable of resolving disputes on the basis of law, of insuring the security of the large and the small, and of creating conditions under which arms can finally be abolished. At the same time we seek to keep peace inside the non-Communist world, where many nations, all of them our friends, are divided over issues which weaken Western unity, which invite Communist intervention, or which threaten to erupt into war. Our efforts in West New Guinea, in the Congo, in the Middle East, and the Indian subcontinent, have been persistent and patient despite criticism from both sides. We have also tried to set an example for others, by seeking to adjust small but significant differences with our own closest neighbors in Mexico and Canada.

Speaking of other nations, I wish to make one point clear. We are bound to many nations by alliances. Those alliances exist because our concern and theirs substantially overlap. Our commitment to defend Western Europe and West Berlin, for example, stands undiminished because of the identity of our vital interests. The United States will make no deal with the Soviet Union at the expense of other nations and other peoples, not merely because they are our partners, but also because their interests and ours converge. Our interests converge, however, not only in defending the frontiers of freedom, but in pursuing the paths of peace. It is our hope, and the purpose of allied policy, to convince the Soviet

Union that she, too, should let each nation choose its own future, so long as that choice does not interfere with the choices of others. The Communist drive to impose their political and economic system on others is the primary cause of world tension today. For there can be no doubt that if all nations could refrain from interfering in the self-determination of others, the peace would be much more assured.

This will require a new effort to achieve world law, a new context for world discussions. It will require increased understanding between the Soviets and ourselves. And increased understanding will require increased contact and communication. One step in this direction is the proposed arrangement for a direct line between Moscow and Washington, to avoid on each side the dangerous delays, misunderstandings, and misreadings of others' actions which might occur at a time of crisis.

We have also been talking in Geneva about our first-step measures of arms controls designed to limit the intensity of the arms race and reduce the risk of accidental war. Our primary long range interest in Geneva, however, is general and complete disarmament, designed to take place by stages, permitting parallel political developments to build the new institutions of peace which would take the place of arms. The pursuit of disarmament has been an effort of this Government since the 1920's. It has been urgently sought by the past three administrations. And however dim the prospects are today, we intend to continue this effort—to continue it in order that all countries, including our own, can better grasp what the problems and possibilities of disarmament are.

The only major area of these negotiations where the end is in sight, yet where a fresh start is badly needed, is in a treaty to outlaw nuclear tests. The conclusion of such a treaty, so near and yet so far, would check the spiraling arms race in one of its most dangerous areas. It would place the nuclear powers in a position to deal more effectively with one of the greatest hazards which man faces in 1963, the further spread of nuclear arms. It would increase our security; it would decrease the prospects of war. Surely this goal is sufficiently important to require our steady pursuit, yielding neither to the temptation to give up the whole effort nor the temptation to give up our insistence on vital and responsible safeguards.

I'm taking this opportunity, therefore, to announce two important decisions in this regard. First, Chairman Khrushchev, Prime Minister Macmillan, and I have agreed that high-level discussions will shortly begin in Moscow looking towards early agreement on a comprehensive test ban treaty. Our hopes must be tempered with the caution of history; but with our hopes go the hopes of all mankind. Second, to make clear our good faith and solemn convictions on this matter, I now declare that the United States does not propose to conduct nuclear tests in the atmosphere so long as other states do not do so. We will not be the first to resume. Such a declaration is no substitute for a formal binding treaty, but I hope it will help us achieve one. Nor would such a treaty be a substitute for disarmament, but I hope it will help us achieve it.

Finally, my fellow Americans, let us examine our attitude towards peace and freedom here at home. The quality and spirit of our own society must justify and support our efforts abroad. We must show it in the dedication of our own lives—as many of you who are graduating today will have an opportunity to do, by serving without pay in the Peace Corps abroad or in the proposed National Service Corps here at home. But wherever we are, we must all, in our daily lives, live up to the age-old faith that peace and freedom walk together. In too many of our cities today, the peace is not secure because freedom is incomplete. It is the responsibility of the executive branch at all levels of government—local, State, and National—to provide and protect that freedom for all of our citizens by all means within our authority. It is the responsibility of the legislative branch at all levels, wherever the authority is not now adequate, to make it adequate. And it is the responsibility of all citizens in all sections of this country to respect the rights of others and respect the law of the land.

All this is not unrelated to world peace. "When a man's ways please the Lord," the Scriptures tell us, "He maketh even his enemies to be at peace with him." And is not peace, in the last analysis, basically a matter of human rights: the right to live out our lives without fear of devastation; the right to breathe air as nature provided it; the right of future generations to a healthy existence?

While we proceed to safeguard our national interests, let us also safeguard human interests. And the elimination of war and arms is clearly in the interest of both. No treaty, however much it may be to the advantage of all, however tightly it may be worded, can provide absolute security against the risks of deception and evasion. But it can, if it is sufficiently effective in its enforcement, and it is sufficiently in the interests of its signers, offer far more security and far fewer risks than an unabated, uncontrolled, unpredictable arms race.

The United States, as the world knows, will never start a war. We do not want a war. We do not now expect a war. This generation of Americans has already had enough—more than enough—of war and hate and oppression.

We shall be prepared if others wish it. We shall be alert to try to stop it. But we shall also do our part to build a world of peace where the weak are safe and the strong are just. We are not helpless before that task or hopeless of its success. Confident and unafraid, we must labor on—not towards a strategy of annihilation but towards a strategy of peace.

SPEECH TO THE IRISH DÁIL

June 28, 1963

———

Mr. Speaker, Prime Minister, Members of the Parliament:

I am grateful for your welcome and for that of your countrymen.

The 13th day of September, 1862, will be a day long remembered in American history. At Fredericksburg, Maryland, thousands of men fought and died on one of the bloodiest battlefields of the American Civil War. One of the most brilliant stories of that day was written by a band of 1200 men who went into battle wearing a green sprig in their hats. They bore a proud heritage and a special courage, given to those who had long fought for the cause of freedom. I am referring, of course, to the Irish Brigade. General Robert E. Lee, the great military leader of the Southern Confederate Forces, said of this group of men after the battle, "The gallant stand which this bold brigade made on the heights of Fredericksburg is well known. Never were men so brave. They ennobled their race by their splendid gallantry on that desperate occasion. Their brilliant though hopeless assaults on our lines excited the hearty applause of our officers and soldiers."

Of the 1200 men who took part in that assault, 280 survived the battle. The Irish Brigade was led into battle on that occasion by Brig. Gen. Thomas F. Meagher, who had participated in the unsuccessful Irish uprising of 1848, was captured by the British and sent in a prison ship to Australia from whence he finally came to America. In the fall of 1862, after serving with distinction and gallantry in some of the toughest fighting of this most bloody struggle, the Irish Brigade was presented with a new set of flags. In the city ceremony, the city chamberlain gave them the motto, "The Union, our Country, and Ireland forever." Their old ones having been torn to shreds in previous battles, Capt. Richard McGee took possession of these flags on December 2d in New York City and arrived with them at the Battle of Fredericksburg and carried them in the battle. Today, in recognition of what these gallant Irishmen and what millions of other Irish have done for my country, and through the generosity of the "Fighting 69th," I would like to present one of these flags to the people of Ireland.

As you can see, gentlemen, the battle honors of the Brigade include Fredericksburg, Chancellorsville, Yorktown, Fair Oaks, Gaines Mill, Allen's Farm, Savage's Station, White Oak Bridge, Glendale, Malvern Hill, Antietam, Gettysburg, and Bristow Station.

I am deeply honored to be your guest in a Free Parliament in a free Ireland. If this nation had achieved its present political and economic stature a century or so ago, my great grandfather might never have left New Ross, and I might, if fortunate, be sitting down there with you. Of course if your own President had never left Brooklyn, he might be standing up here instead of me.

This elegant building, as you know, was once the property of the Fitzgerald family, but I have not come here to claim it. Of all the new relations I have discovered on this trip, I regret to say that no one has yet found any link between me and a great Irish patriot, Lord Edward Fitzgerald. Lord Edward, however, did not like to stay here in his family home because, as he wrote his mother, "Leinster House does not inspire the brightest ideas." That was a long time ago, however. It has also been said by some that a few of the features of this stately mansion served to inspire similar features in the White House in Washington. Whether this is true or not, I know that the White House was designed by James Hoban, a noted Irish-American architect and I have no doubt that he believed by incorporating several features of the Dublin style he would make it more homelike for any President of Irish descent. It was a long wait, but I appreciate his efforts.

There is also an unconfirmed rumor that Hoban was never fully paid for his work on the White House. If this proves to be true, I will speak to our Secretary of the Treasury about it, although I hear his body is not particularly interested in the subject of revenues.

I am proud to be the first American President to visit Ireland during his term of office, proud to be addressing this distinguished assembly, and proud of the welcome you have given me. My presence and your welcome, however, only symbolize the many and the enduring links which have bound the Irish and the Americans since the earliest days.

Benjamin Franklin—the envoy of the American Revolution who was also born in Boston—was received by the Irish Parliament in 1772. It was neither independent nor free from discrimination at the time, but Franklin reported its members "disposed to be friends of America." "By joining our interest with theirs," he said, "a more equitable treatment . . . might be obtained for both nations."

Our interests have been joined ever since. Franklin sent leaflets to Irish freedom fighters. O'Connell was influenced by Washington, and Emmet influenced Lincoln. Irish volunteers played so predominant a role in the American army that Lord Mountjoy lamented in the British Parliament that "we have lost America through the Irish."

John Barry, whose statue we honored yesterday and whose sword is in my office, was only one who fought for liberty in America to set an example for liberty in Ireland. Yesterday was the 117th anniversary of the birth of Charles Stewart Parnell—whose grandfather fought under Barry and whose mother was born in America—and who, at the age of 34, was invited to address the American Congress on the cause of Irish freedom. "I have seen since I have been in this country," he said, "so many tokens of the good wishes of the American people toward Ireland . . ." And today, 83 years later, I can say to you that I have seen in this country so many tokens of good wishes of the Irish people towards America.

And so it is that our two nations, divided by distance, have been united by history. No people ever believed more deeply in the cause of Irish freedom than the people of the United States. And no country contributed more to building my own than your sons and daughters. They came to our shores in a mixture of hope and agony, and I would not underrate the difficulties of their course once they arrived in the United States. They left behind hearts, fields, and a nation yearning to be free. It is no wonder that James Joyce described the Atlantic as a bowl of bitter tears. And an earlier poet wrote, "They are going, going, going, and we cannot bid them stay."

But today this is no longer the country of hunger and famine that those emigrants left behind. It is not rich, and its progress is not yet complete, but it is, according to statistics, one of the best fed countries in the world. Nor is it any longer a country of persecution, political or religious. It is a free country, and that is why any American feels at home.

There are those who regard this history of past strife and exile as better forgotten. But, to use the phrase of Yeats, let us not casually reduce "that great past to a trouble of fools." For we need not feel the bitterness of the past to discover its meaning for the present and the future. And it is the present and the future of Ireland that today holds so much promise to my nation as well as to yours, and, indeed, to all mankind.

For the Ireland of 1963, one of the youngest of nations and oldest of civilizations, has discovered that the achievement of nationhood is not an end but a beginning. In the years since independence, you have undergone a new and peaceful revolution, transforming the face of this land while still holding to the old spiritual and cultural values. You have modernized your economy, harnessed your rivers, diversified your industry, liberalized your trade, electrified your farms, accelerated your rate of growth, and improved the living standards of your people.

The other nations of the world—in whom Ireland has long invested her people and her children—are now investing their capital as well as their vacations here in Ireland. This revolution is not yet over, nor will it be, I am sure, until a fully modern Irish economy shares in world prosperity.

But prosperity is not enough. Eighty-three years ago, Henry Grattan, demanding the more independent Irish Parliament that would always bear his name, denounced those who were satisfied merely by new grants of economic opportunity. "A country," he said, "enlightened as Ireland, chartered as Ireland, armed as Ireland and injured as Ireland will be satisfied with nothing less than liberty." And today, I am certain, free Ireland—a full-fledged member of the world community, where some are not yet free, and where some counsel an acceptance of tyranny—free Ireland will not be satisfied with anything less than liberty.

I am glad, therefore, that Ireland is moving in the mainstream of current world events. For I sincerely believe that your future is as promising as your past is proud, and that your destiny lies not as a peaceful island in a sea of troubles, but as a maker and shaper of world peace.

For self-determination can no longer mean isolation; and the achievement of national independence today means withdrawal from the old status only to return to the world scene with a new one. New nations can build with their former governing powers the same kind of fruitful relationship that Ireland has established with Great Britain—a relationship founded on equality and mutual interests. And no nation, large or small, can be indifferent to the fate of others, near or far. Modern economics, weaponry and communications have made us all realize more than ever that we are one human family and this one planet is our home.

"The world is large," wrote John Boyle O'Reilly.

> *The world is large when its weary*
> *leagues two loving hearts divide,*
> *But the world is small when your enemy*
> *is loose on the other side.*

The world is even smaller today, though the enemy of John Boyle O'Reilly is no longer a hostile power. Indeed, across the gulfs and barriers that now divide us, we must remember that there are no permanent enemies. Hostility today is a fact, but it is not a ruling law. The supreme reality of our time is our indivisibility as children of God and our common vulnerability on this planet.

Some may say that all this means little to Ireland. In an age when "history moves with the tramp of earthquake feet"—in an age when a handful of men and nations have the power literally to devastate mankind—in an age when the needs of the developing nations are so staggering that even the richest lands often groan with the burden of assistance—in such an age, it may be asked, how can a nation as small as Ireland play much of a role on the world stage?

I would remind those who ask that question, including those in other small countries, of the words of one of the great orators of the English language:

"All the world owes much to the little 'five feet high' nations. The greatest art of the world was the work of little nations. The most enduring literature of the world came from little nations. The heroic deeds that thrill humanity through generations were the deeds of little nations fighting for their freedom. And oh, yes, the salvation of mankind came through a little nation."

Ireland has already set an example and a standard for other small nations to follow.

This has never been a rich or powerful country, and yet, since earliest times, its influence on the world has been rich and powerful. No larger nation did more to keep Christianity and Western culture alive in their darkest centuries. No larger nation did more to spark the cause of independence in America, indeed, around the world. And no larger nation has ever provided the world with more literary and artistic genius.

This is an extraordinary country. George Bernard Shaw, speaking as an Irishman, summed up an approach to life: Other people, he said "see things and . . . say 'Why?' . . . But I dream things that never were—and I say: 'Why not?' "

It is that quality of the Irish—that remarkable combination of hope, confidence and imagination—that is needed more than ever today. The problems of the world cannot possibly be solved by skeptics or cynics, whose horizons are limited by the obvious realities. We need men who can dream of things that never were, and ask why not. It matters not how small a nation is that seeks world peace and freedom, for, to paraphrase a citizen of my country, "the humblest nation of all the world, when clad in the armor of a righteous cause, is stronger than all the hosts of Error."

Ireland is clad in the cause of national and human liberty with peace. To the extent that the peace is disturbed by conflict between the former colonial powers and the new and developing nations, Ireland's role is unique. For every new nation knows that Ireland was the first of the small nations in the 20th century to win its struggle for independence, and that the Irish have traditionally sent their doctors and technicians and soldiers and priests to help other lands to keep their liberty alive.

At the same time, Ireland is part of Europe, associated with the Council of Europe, progressing in the context of Europe, and a prospective member of an expanded European Common Market. Thus Ireland has excellent relations with both the new and the old, the confidence of both sides and an opportunity to act where the actions of greater powers might be looked upon with suspicion.

The central issue of freedom, however, is between those who believe in self-determination and those in the East who would impose on others the harsh and oppressive Communist system; and here your nation wisely rejects the role of a go-between or a mediator. Ireland pursues an independent course in foreign policy, but it is not neutral between liberty and tyranny and never will be.

For knowing the meaning of foreign domination, Ireland is the example and inspiration to those enduring endless years of oppression. It was fitting and appropriate that this nation played a leading role in censuring the suppression of the Hungarian revolution, for how many times was Ireland's quest for freedom suppressed only to have that quest renewed by the succeeding generation? Those who suffer beyond that wall I saw on Wednesday in Berlin must not despair of their future. Let them remember the constancy, the faith, the endurance, and the final success of the Irish. And let them remember, as I heard sung by your sons and daughters yesterday in Wexford, the words, "the boys of Wexford, who fought with heart and hand, to burst in twain the galling chain and free our native land."

The major forum for your nation's greater role in world affairs is that of protector of the weak and voice of the small, the United Nations. From Cork to the Congo, from Galway to the Gaza Strip, from this legislative assembly to the United Nations, Ireland is sending its most talented men to do the world's most important work—the work of peace.

In a sense, this export of talent is in keeping with an historic

Irish role—but you no longer go as exiles and emigrants but for the service of your country and, indeed, of all men. Like the Irish missionaries of medieval days, like the "wild geese" after the Battle of the Boyne, you are not content to sit by your fireside while others are in need of your help. Nor are you content with the recollections of the past when you face the responsibilities of the present.

Twenty-six sons of Ireland have died in the Congo; many others have been wounded. I pay tribute to them and to all of you for your commitment and dedication to world order. And their sacrifice reminds us all that we must not falter now.

The United Nations must be fully and fairly financed. Its peace-keeping machinery must be strengthened. Its institutions must be developed until some day, and perhaps some distant day, a world of law is achieved.

Ireland's influence in the United Nations is far greater than your relative size. You have not hesitated to take the lead on such sensitive issues as the Kashmir dispute. And you sponsored that most vital resolution, adopted by the General Assembly, which opposed the spread of nuclear arms to any nation not now possessing them, urging an international agreement with inspection and controls. And I pledge to you that the United States of America will do all in its power to achieve such an agreement and fulfill your resolution.

I speak of these matters today—not because Ireland is unaware of its role—but I think it important that you know that we know what you have done. And I speak to remind the other small nations that they, too, can and must help build a world peace. They, too, as we all are, are dependent on the United Nations for security, for an equal chance to be heard, for progress towards a world made safe for diversity.

The peace-keeping machinery of the United Nations cannot work without the help of the smaller nations, nations whose forces threaten no one and whose forces can thus help create a world in which no nation is threatened. Great powers have their responsibilities and their burdens, but the smaller nations of the world must fulfill their obligations as well.

A great Irish poet once wrote: "I believe profoundly . . . in the future of Ireland . . . that this is an isle of destiny, that that destiny will be glorious . . . and that when our hour is come, we will have something to give to the world."

My friends: Ireland's hour has come. You have something to give to the world—and that is a future of peace with freedom.

Thank you.

ADDRESS TO THE NATION
ON THE PARTIAL NUCLEAR
TEST BAN TREATY

July 26, 1963

————

Good evening, my fellow citizens:

I speak to you tonight in a spirit of hope. Eighteen years ago the advent of nuclear weapons changed the course of the world as well as the war. Since that time, all mankind has been struggling to escape from the darkening prospect of mass destruction on earth. In an age when both sides have come to possess enough nuclear power to destroy the human race several times over, the world of communism and the world of free choice have been caught up in a vicious circle of conflicting ideology and interest. Each increase of tension has produced an increase of arms; each increase of arms has produced an increase of tension.

In these years, the United States and the Soviet Union have frequently communicated suspicion and warnings to each other, but very rarely hope. Our representatives have met at the summit and at the brink; they have met in Washington and in Moscow; in Geneva and at the United Nations. But too often these meetings have produced only darkness, discord, or disillusion.

Yesterday a shaft of light cut into the darkness. Negotiations were concluded in Moscow on a treaty to ban all nuclear tests in the atmosphere, in outer space, and under water. For the first time, an agreement has been reached on bringing the forces of nuclear destruction under international control—a goal first sought in 1946 when Bernard Baruch presented a comprehensive control plan to the United Nations.

That plan, and many subsequent disarmament plans, large and small, have all been blocked by those opposed to international inspection. A ban on nuclear tests, however, requires on-the-spot inspection only for underground tests. This Nation now possesses a variety of techniques to detect the nuclear tests of other nations which are conducted in the air or under water, for such tests produce unmistakable signs which our modern instruments can pick up.

The treaty initialed yesterday, therefore, is a limited treaty which permits continued underground testing and prohibits only those tests that we ourselves can police. It requires no control posts, no onsite inspection, no international body.

We should also understand that it has other limits as well. Any nation which signs the treaty will have an opportunity to withdraw if it finds that extraordinary events related to the subject matter of the treaty have jeopardized its supreme interests; and no nation's right of self-defense will in any way be impaired. Nor does this treaty mean an end to the threat of nuclear war. It will not reduce nuclear stockpiles; it will not halt the production of nuclear weapons; it will not restrict their use in time of war.

Nevertheless, this limited treaty will radically reduce the nuclear testing which would otherwise be conducted on both sides; it will prohibit the United States, the United Kingdom, the Soviet Union, and all others who sign it, from engaging in the atmospheric tests which have so alarmed mankind; and it offers to all the world a welcome sign of hope.

For this is not a unilateral moratorium, but a specific and solemn legal obligation. While it will not prevent this Nation from testing underground, or from being ready to conduct atmospheric tests if the acts of others so require, it gives us a concrete opportunity to extend its coverage to other nations and later to other forms of nuclear tests.

This treaty is in part the product of Western patience and vigilance. We have made clear—most recently in Berlin and Cuba—our deep resolve to protect our security and our freedom against any form of aggression. We have also made clear our steadfast determination to limit the arms race. In three administrations, our soldiers and diplomats have worked together to this end, always supported by Great Britain. Prime Minister Macmillan joined with President Eisenhower in proposing a limited test ban in 1959, and again with me in 1961 and 1962.

But the achievement of this goal is not a victory for one side—it is a victory for mankind. It reflects no concessions either to or by the Soviet Union. It reflects simply our common recognition of the dangers in further testing.

This treaty is not the millennium. It will not resolve all conflicts, or cause the Communists to forego their ambitions, or eliminate the dangers of war. It will not reduce our need for arms or allies or programs of assistance to others. But it is an important first step— a step towards peace—a step towards reason—a step away from war.

Here is what this step can mean to you and to your children and your neighbors:

First, this treaty can be a step towards reduced world tension and broader areas of agreement. The Moscow talks have reached no agreement on any other subject, nor is this treaty conditioned on any other matter. Under Secretary Harriman made it clear that any nonaggression arrangements across the division in Europe would require full consultation with our allies and full attention to their interests. He also made clear our strong preference for a more comprehensive treaty banning all tests everywhere, and our ultimate hope for general and complete disarmament. The Soviet Government, however, is still unwilling to accept the inspection such goals require.

No one can predict with certainty, therefore, what further agreements, if any, can be built on the foundations of this one. They could include controls on preparations for surprise attack, or on numbers and type of armaments. There could be further limitations on the spread of nuclear weapons. The important point is that efforts to seek new agreements will go forward.

But the difficulty of predicting the next step is no reason to be reluctant about this step. Nuclear test ban negotiations have long been a symbol of East-West disagreement. If this treaty can also be a symbol—if it can symbolize the end of one era and the beginning of another—if both sides can by this treaty gain confidence and experience in peaceful collaboration—then this short and simple treaty may well become an historic mark in man's age-old pursuit of peace.

Western policies have long been designed to persuade the Soviet Union to renounce aggression, direct or indirect, so that their people and all people may live and let live in peace. The unlimited testing of new weapons of war cannot lead towards that end—but this treaty, if it can be followed by further progress, can clearly move in that direction.

I do not say that a world without aggression or threats of war would be an easy world. It will bring new problems, new challenges from the Communists, new dangers of relaxing our vigilance or of mistaking their intent.

But those dangers pale in comparison to those of the spiraling arms race and a collision course towards war. Since the beginning of history, war has been mankind's constant companion. It has been the rule, not the exception. Even a nation as young and as peace-loving as our own has fought through eight wars. And three times in the last two years and a half I have been required to report to you as President that this Nation and the Soviet Union stood on the verge of direct military confrontation—in Laos, in Berlin, and in Cuba.

A war today or tomorrow, if it led to nuclear war, would not be like any war in history. A full-scale nuclear exchange, lasting less than 60 minutes, with the weapons now in existence, could wipe out more than 300 million Americans, Europeans, and Russians, as well as untold numbers elsewhere. And the survivors, as Chairman Khrushchev warned the Communist Chinese, "the survivors would envy the dead." For they would inherit a world so devastated by explosions and poison and fire that today we cannot even conceive of its horrors. So let us try to turn the world away from war. Let us make the most of this opportunity, and every opportunity, to reduce tension, to slow down the perilous nuclear arms race, and to check the world's slide toward final annihilation.

Second, this treaty can be a step towards freeing the world from the fears and dangers of radioactive fallout. Our own atmospheric tests last year were conducted under conditions which restricted such fallout to an absolute minimum. But over the years the number and the yield of weapons tested have rapidly increased and so have the radioactive hazards from such testing. Continued unrestricted testing by the nuclear powers, joined in time by other nations which may be less adept in limiting pollution, will increasingly contaminate the air that all of us must breathe.

Even then, the number of children and grandchildren with cancer in their bones, with leukemia in their blood, or with poison in their lungs might seem statistically small to some, in comparison with natural health hazards. But this is not a natural health hazard—and it is not a statistical issue. The loss of even one human life, or the malformation of even one baby—who may be born long after we are gone—should be of concern to us all. Our children and grandchildren are not merely statistics toward which we can be indifferent.

Nor does this affect the nuclear powers alone. These tests befoul the air of all men and all nations, the committed and the uncommitted alike, without their knowledge and without their consent. That is why the continuation of atmospheric testing causes so many countries to regard all nuclear powers as equally evil; and we can hope that its prevention will enable those countries to see the world more clearly, while enabling all the world to breathe more easily.

Third, this treaty can be a step toward preventing the spread of nuclear weapons to nations not now possessing them. During the next several years, in addition to the four current nuclear powers, a small but significant number of nations will have the intellectual, physical, and financial resources to produce both nuclear weapons and the means of delivering them. In time, it is estimated, many other nations will have either this capacity or other ways of obtaining nuclear warheads, even as missiles can be commercially purchased today.

I ask you to stop and think for a moment what it would mean to have nuclear weapons in so many hands, in the hands of countries large and small, stable and unstable, responsible and irresponsible, scattered throughout the world. There would be no rest for anyone then, no stability, no real security, and no chance of effective disarmament. There would only be the increased chance of accidental war, and an increased necessity for the great powers to involve themselves in what otherwise would be local conflicts.

If only one thermonuclear bomb were to be dropped on any American, Russian, or any other city, whether it was launched by accident or design, by a madman or by an enemy, by a large nation or by a small, from any corner of the world, that one bomb could release more destructive power on the inhabitants of that one helpless city than all the bombs dropped in the Second World War.

Neither the United States nor the Soviet Union nor the United Kingdom nor France can look forward to that day with equanimity. We have a great obligation, all four nuclear powers have a great obligation, to use whatever time remains to prevent the spread of nuclear weapons, to persuade other countries not to test, transfer, acquire, possess, or produce such weapons.

This treaty can be the opening wedge in that campaign. It provides that none of the parties will assist other nations to test in the forbidden environments. It opens the door for further agreements on the control of nuclear weapons, and it is open for all nations to sign, for it is in the interest of all nations, and already we have heard from a number of countries who wish to join with us promptly.

Fourth and finally, this treaty can limit the nuclear arms race in ways which, on balance, will strengthen our Nation's security far more than the continuation of unrestricted testing. For in today's world, a nation's security does not always increase as its arms increase, when its adversary is doing the same, and unlimited competition in the testing and development of new types of destructive nuclear weapons will not make the world safer for either side. Under this limited treaty, on the other hand, the testing of other nations could never be sufficient to offset the ability of our strategic forces to deter or survive a nuclear attack and to penetrate and destroy an aggressor's homeland.

We have, and under this treaty we will continue to have, the nuclear strength that we need. It is true that the Soviets have tested nuclear weapons of a yield higher than that which we thought to be necessary, but the hundred megaton bomb of which they spoke two years ago does not and will not change the balance of strategic power. The United States has chosen, deliberately, to concentrate on more mobile and more efficient weapons, with lower but entirely sufficient yield, and our security is, therefore, not impaired by the treaty I am discussing.

It is also true, as Mr. Khrushchev would agree, that nations cannot afford in these matters to rely simply on the good faith of their adversaries. We have not, therefore, overlooked the risk of secret violations. There is at present a possibility that deep in outer space, that hundreds and thousands and millions of miles away from the earth illegal tests might go undetected. But we already have the capability to construct a system of observation that would make such tests almost impossible to conceal, and we can decide at any time whether such a system is needed in the light of the limited risk to us and the limited reward to others of violations attempted at that range. For any tests which might be conducted so far out in space, which cannot be conducted more easily and efficiently and legally underground, would necessarily be of such a magnitude that they would be extremely difficult to conceal. We can also employ new devices to check on the testing of smaller weapons in the lower atmosphere. Any violation, moreover, involves, along with the risk of detection, the end of the treaty and the worldwide consequences for the violator.

Secret violations are possible and secret preparations for a sudden withdrawal are possible, and thus our own vigilance and strength must be maintained, as we remain ready to withdraw and to resume all forms of testing, if we must. But it would be a mistake to assume that this treaty will be quickly broken. The gains of illegal testing are obviously slight compared to their cost, and the hazard of discovery, and the nations which have initialed and will sign this treaty prefer it, in my judgment, to unrestricted testing as a matter of their own self-interests, for these nations, too, and all nations, have a stake in limiting the arms race, in holding the spread of nuclear weapons, and in breathing air that is not radioactive. While it may be theoretically possible to demonstrate the risks inherent in any treaty, and such risks in this treaty are small, the far greater risks to our security are the risks of unrestricted testing, the risk of a nuclear arms race, the risk of new nuclear powers, nuclear pollution, and nuclear war.

This limited test ban, in our most careful judgment, is safer by far for the United States than an unlimited nuclear arms race. For all these reasons, I am hopeful that this Nation will promptly approve the limited test ban treaty. There will, of course, be debate in the country and in the Senate. The Constitution wisely requires the advice and consent of the Senate to all treaties, and that consultation has already begun. All this is as it should be. A document which may mark an historic and constructive opportunity for the world deserves an historic and constructive debate.

It is my hope that all of you will take part in that debate, for this treaty is for all of us. It is particularly for our children and our grandchildren, and they have no lobby here in Washington. This debate will involve military, scientific, and political experts, but it must be not left to them alone. The right and the responsibility are yours.

If we are to open new doorways to peace, if we are to seize this rare opportunity for progress, if we are to be as bold and farsighted in our control of weapons as we have been in their invention, then let us now show all the world on this side of the wall and the other that a strong America also stands for peace. There is no cause for complacency.

We have learned in times past that the spirit of one moment or place can be gone in the next. We have been disappointed more than once, and we have no illusions now that there are shortcuts on the road to peace. At many points around the globe the Communists are continuing their efforts to exploit weakness and poverty. Their concentration of nuclear and conventional arms must still be deterred.

The familiar contest between choice and coercion, the familiar places of danger and conflict, are all still there, in Cuba, in Southeast Asia, in Berlin, and all around the globe, still requiring all the strength and the vigilance that we can muster. Nothing could more greatly damage our cause than if we and our allies were to believe that peace has already been achieved, and that our strength and unity were no longer required.

But now, for the first time in many years, the path of peace may be open. No one can be certain what the future will bring. No one can say whether the time has come for an easing of the struggle. But history and our own conscience will judge us harsher if we do not now make every effort to test our hopes by action, and this is the place to begin.

According to the ancient Chinese proverb, "A journey of a thousand miles must begin with a single step."

My fellow Americans, let us take that first step. Let us, if we can, step back from the shadows of war and seek out the way of peace. And if that journey is a thousand miles, or even more, let history record that we, in this land, at this time, took the first step.

Thank you and good night.

SPEECH TO THE 18TH GENERAL ASSEMBLY OF THE UNITED NATIONS

September 20, 1963

———

Mr. President—as one who has taken some interest in the election of Presidents, I want to congratulate you on your election to this high office—Mr. Secretary General, delegates to the United Nations, ladies and gentlemen:

We meet again in the quest for peace.

Twenty-four months ago, when I last had the honor of addressing this body, the shadow of fear lay darkly across the world. The freedom of West Berlin was in immediate peril. Agreement on a neutral Laos seemed remote. The mandate of the United Nations in the Congo was under fire. The financial outlook for this organization was in doubt. Dag Hammarskjold was dead. The doctrine of troika was being pressed in his place, and atmospheric tests had been resumed by the Soviet Union.

Those were anxious days for mankind—and some men wondered aloud whether this organization could survive. But the 16th and 17th General Assemblies achieved not only survival but progress. Rising to its responsibility, the United Nations helped reduce the tensions and helped to hold back the darkness.

Today the clouds have lifted a little so that new rays of hope can break through. The pressures on West Berlin appear to be temporarily eased. Political unity in the Congo has been largely restored. A neutral coalition in Laos, while still in difficulty, is at least in being. The integrity of the United Nations Secretariat has been reaffirmed. A United Nations Decade of Development is under way. And, for the first time in 17 years of effort, a specific step has been taken to limit the nuclear arms race.

I refer, of course, to the treaty to ban nuclear tests in the atmosphere, outer space, and under water—concluded by the Soviet Union, the United Kingdom, and the United States—and already signed by nearly 100 countries. It has been hailed by people the world over who are thankful to be free from the fears of nuclear fallout, and I am confident that on next Tuesday at 10:30 o'clock in the morning it will receive the overwhelming endorsement of the Senate of the United States.

The world has not escaped from the darkness. The long shadows of conflict and crisis envelop us still. But we meet today in an atmosphere of rising hope, and at a moment of comparative calm. My presence here today is not a sign of crisis, but of confidence. I am not here to report on a new threat to the peace or new signs of war. I have come to salute the United Nations and to show the support of the American people for your daily deliberations.

For the value of this body's work is not dependent on the existence of emergencies—nor can the winning of peace consist only of dramatic victories. Peace is a daily, a weekly, a monthly process, gradually changing opinions, slowly eroding old barriers, quietly building new structures. And however undramatic the pursuit of peace, that pursuit must go on.

Today we may have reached a pause in the cold war—but that is not a lasting peace. A test ban treaty is a milestone—but it is not the millennium. We have not been released from our obligations—

we have been given an opportunity. And if we fail to make the most of this moment and this momentum—if we convert our new-found hopes and understandings into new walls and weapons of hostility—if this pause in the cold war merely leads to its renewal and not to its end—then the indictment of posterity will rightly point its finger at us all. But if we can stretch this pause into a period of cooperation—if both sides can now gain new confidence and experience in concrete collaborations for peace—if we can now be as bold and farsighted in the control of deadly weapons as we have been in their creation—then surely this first small step can be the start of a long and fruitful journey.

The task of building the peace lies with the leaders of every na-tion, large and small. For the great powers have no monopoly on conflict or ambition. The cold war is not the only expression of tension in this world—and the nuclear race is not the only arms race. Even little wars are dangerous in a nuclear world. The long labor of peace is an undertaking for every nation—and in this ef-fort none of us can remain unaligned. To this goal none can be uncommitted.

The reduction of global tension must not be an excuse for the narrow pursuit of self-interest. If the Soviet Union and the United States, with all of their global interests and clashing commitments of ideology, and with nuclear weapons still aimed at each other today, can find areas of common interest and agreement, then surely other nations can do the same—nations caught in regional conflicts, in racial issues, or in the death throes of old colonialism. Chronic disputes which divert precious resources from the needs of the people or drain the energies of both sides serve the interests of no one—and the badge of responsibility in the modern world is a willingness to seek peaceful solutions.

It is never too early to try; and it's never too late to talk; and it's high time that many disputes on the agenda of this Assembly were taken off the debating schedule and placed on the negotiating table.

The fact remains that the United States, as a major nuclear

power, does have a special responsibility in the world. It is, in fact, a threefold responsibility—a responsibility to our own citizens; a responsibility to the people of the whole world who are affected by our decisions; and to the next generation of humanity. We believe the Soviet Union also has these special responsibilities—and that those responsibilities require our two nations to concentrate less on our differences and more on the means of resolving them peacefully. For too long both of us have increased our military budgets, our nuclear stockpiles, and our capacity to destroy all life on this hemisphere—human, animal, vegetable—without any corresponding increase in our security.

Our conflicts, to be sure, are real. Our concepts of the world are different. No service is performed by failing to make clear our disagreements. A central difference is the belief of the American people in the self-determination of all people.

We believe that the people of Germany and Berlin must be free to reunite their capital and their country.

We believe that the people of Cuba must be free to secure the fruits of the revolution that have been betrayed from within and exploited from without.

In short, we believe that all the world—in Eastern Europe as well as Western, in Southern Africa as well as Northern, in old nations as well as new—that people must be free to choose their own future, without discrimination or dictation, without coercion or subversion.

These are the basic differences between the Soviet Union and the United States, and they cannot be concealed. So long as they exist, they set limits to agreement, and they forbid the relaxation of our vigilance. Our defense around the world will be maintained for the protection of freedom—and our determination to safeguard that freedom will measure up to any threat or challenge.

But I would say to the leaders of the Soviet Union, and to their people, that if either of our countries is to be fully secure, we need a much better weapon than the H-bomb—a weapon better than ballistic missiles or nuclear submarines—and that better weapon is peaceful cooperation.

We have, in recent years, agreed on a limited test ban treaty, on an emergency communications link between our capitals, on a statement of principles for disarmament, on an increase in cultural exchange, on cooperation in outer space, on the peaceful exploration of the Antarctic, and on tempering last year's crisis over Cuba.

I believe, therefore, that the Soviet Union and the United States, together with their allies, can achieve further agreements— agreements which spring from our mutual interest in avoiding mutual destruction.

There can be no doubt about the agenda of further steps. We must continue to seek agreements on measures which prevent war by accident or miscalculation. We must continue to seek agreements on safeguards against surprise attack, including observation posts at key points. We must continue to seek agreement on further measures to curb the nuclear arms race, by controlling the transfer of nuclear weapons, converting fissionable materials to peaceful purposes, and banning underground testing, with adequate inspection and enforcement. We must continue to seek agreement on a freer flow of information and people from East to West and West to East.

We must continue to seek agreement, encouraged by yesterday's affirmative response to this proposal by the Soviet Foreign Minister, on an arrangement to keep weapons of mass destruction out of outer space. Let us get our negotiators back to the negotiating table to work out a practicable arrangement to this end.

In these and other ways, let us move up the steep and difficult path toward comprehensive disarmament, securing mutual confidence through mutual verification, and building the institutions of peace as we dismantle the engines of war. We must not let failure to agree on all points delay agreements where agreement is possible. And we must not put forward proposals for propaganda purposes.

Finally, in a field where the United States and the Soviet Union have a special capacity—in the field of space—there is room for new cooperation, for further joint efforts in the regulation and exploration of space. I include among these possibilities a joint expedition to the moon. Space offers no problems of sovereignty; by resolution of this Assembly, the members of the United Nations have foresworn any claim to territorial rights in outer space or on celestial bodies, and declared that international law and the United Nations Charter will apply. Why, therefore, should man's first flight to the moon be a matter of national competition? Why should the United States and the Soviet Union, in preparing for such expeditions, become involved in immense duplications of research, construction, and expenditure? Surely we should explore whether the scientists and astronauts of our two countries—indeed of all the world—cannot work together in the conquest of space, sending some day in this decade to the moon not the representatives of a single nation, but the representatives of all of our countries.

All these and other new steps toward peaceful cooperation may be possible. Most of them will require on our part full consultation with our allies—for their interests are as much involved as our own, and we will not make an agreement at their expense. Most of them will require long and careful negotiation. And most of them will require a new approach to the cold war—a desire not to "bury" one's adversary, but to compete in a host of peaceful arenas, in ideas, in production, and ultimately in service to all mankind.

The contest will continue—the contest between those who see a monolithic world and those who believe in diversity—but it should be a contest in leadership and responsibility instead of destruction, a contest in achievement instead of intimidation. Speaking for the United States of America, I welcome such a contest. For we believe that truth is stronger than error—and that freedom is more enduring than coercion. And in the contest for a better life, all the world can be a winner.

The effort to improve the conditions of man, however, is not a task for the few. It is the task of all nations—acting alone, acting in groups, acting in the United Nations, for plague and pestilence, and plunder and pollution, the hazards of nature, and the hunger of children are the foes of every nation. The earth, the sea, and the air are the concern of every nation. And science, technology, and education can be the ally of every nation.

Never before has man had such capacity to control his own environment, to end thirst and hunger, to conquer poverty and disease, to banish illiteracy and massive human misery. We have the power to make this the best generation of mankind in the history of the world—or to make it the last.

The United States since the close of the war has sent over $100 billion worth of assistance to nations seeking economic viability. And 2 years ago this week we formed a Peace Corps to help interested nations meet the demand for trained manpower. Other industrialized nations whose economies were rebuilt not so long ago with some help from us are now in turn recognizing their responsibility to the less developed nations.

The provision of development assistance by individual nations must go on. But the United Nations also must play a larger role in helping bring to all men the fruits of modern science and industry. A United Nations conference on this subject held earlier this year in Geneva opened new vistas for the developing countries. Next year a United Nations Conference on Trade will consider the needs of these nations for new markets. And more than four-fifths of the entire United Nations system can be found today mobilizing the weapons of science and technology for the United Nations' Decade of Development.

But more can be done.

- A world center for health communications under the World Health Organization could warn of epidemics and the adverse effects of certain drugs as well as transmit the results of new experiments and new discoveries.
- Regional research centers could advance our common medical knowledge and train new scientists and doctors for new nations.
- A global system of satellites could provide communication and weather information for all corners of the earth.
- A worldwide program of conservation could protect the forest and wild game preserves now in danger of extinction for all time, improve the marine harvest of food from our oceans, and prevent the contamination of air and water by industrial as well as nuclear pollution.
- And, finally, a worldwide program of farm productivity and food distribution, similar to our country's "Food for Peace" program, could now give every child the food he needs.

But man does not live by bread alone—and the members of this organization are committed by the Charter to promote and respect human rights. Those rights are not respected when a Buddhist priest is driven from his pagoda, when a synagogue is shut down, when a Protestant church cannot open a mission, when a Cardinal is forced into hiding, or when a crowded church service is bombed. The United States of America is opposed to discrimination and persecution on grounds of race and religion anywhere in the world, including our own Nation. We are working to right the wrongs of our own country.

Through legislation and administrative action, through moral and legal commitment, this Government has launched a determined effort to rid our Nation of discrimination which has existed far too long—in education, in housing, in transportation, in employment, in the civil service, in recreation, and in places of public accommodation. And therefore, in this or any other forum, we do not hesitate to condemn racial or religious injustice, whether committed or permitted by friend or foe.

I know that some of you have experienced discrimination in this country. But I ask you to believe me when I tell you that this is not the wish of most Americans—that we share your regret and resentment—and that we intend to end such practices for all time to come, not only for our visitors, but for our own citizens as well.

I hope that not only our Nation but all other multiracial societies will meet these standards of fairness and justice. We are opposed to apartheid and all forms of human oppression. We do not advocate the rights of black Africans in order to drive out white Africans. Our concern is the right of all men to equal protection under the law—and since human rights are indivisible, this body cannot stand aside when those rights are abused and neglected by any member state.

New efforts are needed if this Assembly's Declaration of Human Rights, now 15 years old, is to have full meaning. And new means should be found for promoting the free expression and trade of ideas—through travel and communication, and through increased exchanges of people, and books, and broadcasts. For as the world renounces the competition of weapons, competition in ideas must flourish—and that competition must be as full and as fair as possible.

The United States delegation will be prepared to suggest to the United Nations initiatives in the pursuit of all the goals. For this is an organization for peace—and peace cannot come without work and without progress.

The peacekeeping record of the United Nations has been a proud one, though its tasks are always formidable. We are fortunate to have the skills of our distinguished Secretary General and the brave efforts of those who have been serving the cause of peace in the Congo, in the Middle East, in Korea and Kashmir, in West New Guinea and Malaysia. But what the United Nations has done in the past is less important than the tasks for the future. We cannot take its peacekeeping machinery for granted. That machinery must be soundly financed—which it cannot be if some members are allowed to prevent it from meeting its obligations by failing to meet their own. The United Nations must be supported by all those who exercise their franchise here. And its operations must be backed to the end.

Too often a project is undertaken in the excitement of a crisis and then it begins to lose its appeal as the problems drag on and the bills pile up. But we must have the steadfastness to see every enterprise through.

It is, for example, most important not to jeopardize the extraordinary United Nations gains in the Congo. The nation which sought this organization's help only 3 years ago has now asked the United Nations' presence to remain a little longer. I believe this Assembly should do what is necessary to preserve the gains already made and to protect the new nation in its struggle for progress. Let us complete what we have started. For "No man who puts his hand to the plow and looks back," as the Scriptures tell us, "No man who puts his hand to the plow and looks back is fit for the Kingdom of God."

I also hope that the recent initiative of several members in preparing standby peace forces for United Nations call will encourage similar commitments by others. This Nation remains ready to provide logistic and other material support.

Policing, moreover, is not enough without provision for pacific settlement. We should increase the resort to special missions of fact-finding and conciliation, make greater use of the International Court of Justice, and accelerate the work of the International Law Commission.

The United Nations cannot survive as a static organization. Its obligations are increasing as well as its size. Its Charter must be changed as well as its customs. The authors of that Charter did not intend that it be frozen in perpetuity. The science of weapons and war has made us all, far more than 18 years ago in San Francisco, one world and one human race, with one common destiny. In such a world, absolute sovereignty no longer assures us of absolute security. The conventions of peace must pull abreast and then ahead of the inventions of war. The United Nations, building on its successes and learning from its failures, must be developed into a genuine world security system.

But peace does not rest in charters and covenants alone. It lies in the hearts and minds of all people. And if it is cast out there, then no act, no pact, no treaty, no organization can hope to preserve it without the support and the wholehearted commitment of all people. So let us not rest all our hopes on parchment and on paper; let us strive to build peace, a desire for peace, a willingness to work for peace, in the hearts and minds of all our people. I believe that we can. I believe the problems of human destiny are not beyond the reach of human beings.

Two years ago I told this body that the United States had proposed, and was willing to sign, a limited test ban treaty. Today that treaty has been signed. It will not put an end to war. It will not remove basic conflicts. It will not secure freedom for all. But it can be a lever, and Archimedes, in explaining the principles of the lever, was said to have declared to his friends: "Give me a place where I can stand—and I shall move the world."

My fellow inhabitants of this planet: Let us take our stand here in this Assembly of nations. And let us see if we, in our own time, can move the world to a just and lasting peace.

ACKNOWLEDGMENTS

I am grateful for the support and confidence that many people have given to me as I undertook this project. They might easily have dissuaded me from treading on the historian's territory, but my friends and colleagues knew how much I have been moved by JFK's words and they offered me help and encouragement rather than skepticism. I deeply hope that their support has been justified by the outcome. Of course, especially in this context, all limitations of this work are strictly my own.

My greatest thanks go to Claire Bulger, my special assistant and an indefatigable, incisive, and unerringly supportive researcher, adviser, and editor throughout the project. Claire uncovered wonderful documents in the Kennedy Library, chased down facts and references, and assisted in the completion of the manuscript in all ways. Thanks as well to all the helpful staff at the Kennedy Library, especially Stephen Plotkin. My former special assistant, Aniket Shah, also generously contributed his time and advice to this project, and his careful reading of the manuscript helped to bring it to life.

I am of course grateful to the late Theodore Sorensen, most of all for his inspiration and vast contribution to our world, but also for having the confidence that I could write a book on his favorite of all JFK speeches. That encouragement propelled me throughout the project, as did the warmth and continuing assistance of Gillian Sorensen, Ted's wonderful wife and senior UN diplomat.

Many great scholars helped me to understand Kennedy's peace initiative. I am especially thankful to Professor Marc Trachtenberg, whose wise and penetrating scholarship on the Cold War in *A Constructed Peace* and other works deeply informed my understanding of Kennedy's challenges and triumphs. Professor Trachtenberg kindly gave suggestions on an early draft of the manuscript. I also thank Professor Amitai Etzioni, a longtime champion of peace and psychological interpreter of foreign affairs, whose seminal work *The Hard Way to Peace* deepened my understanding of the Cold War and the reasons for Kennedy's approach and success. My colleague Professor Richard Gardner, leading diplomat and State Department official in the Kennedy administration, generously offered his important insights and advice.

My wife, Sonia, children Lisa, Adam, and Hannah, and son-in-law, Matt, as always, have been intimately engaged in this work at every stage. They have been pressed into service in countless ways: listening with me on endless occasions to the recorded speeches; discussing their meaning and beauty; hearing my theories on this or that aspect of JFK's presidency; reading the manuscript; and of course suffering the long hours of my diverted attention in a house strewn with books and papers. I especially thank Adam, a writer and thinker of exceptional brilliance, for helping me at every turn of the writing. His contributions are reflected throughout the manuscript, though all clumsy phrasing that remains is surely my own.

As always, my literary agent, Andrew Wylie, and Random

House editor, Jonathan Jao, have enabled me to give life to an idea. Andrew's support and encouragement, and Jonathan's trenchant advice and editing, make my book writing possible and a joyful experience for me. I also thank all of the remarkable professionals at Random House for their care, advice, and support.

BIBLIOGRAPHY

Allison, Graham T., and Philip Zelikow. *Essence of Decision: Explaining the Cuban Missile Crisis*. 2nd ed. New York: Longman, 1999.

Axelrod, Robert M. *The Evolution of Cooperation*. New York: Basic Books, 2006.

Beschloss, Michael R. *The Crisis Years: Kennedy and Khrushchev, 1960–1963*. New York: Edward Burlingame Books, 1991.

———. *MAYDAY: Eisenhower, Khrushchev, and the U-2 Affair*. New York: Harper & Row, 1986.

Cooper, John Milton, Jr. *Woodrow Wilson*. New York: Random House, 2009.

Cousins, Norman. *The Improbable Triumvirate: John F. Kennedy, Pope John, Nikita Khrushchev*. New York: W. W. Norton, 1972.

Dallek, Robert. *An Unfinished Life: John F. Kennedy, 1917–1963*. Boston: Little, Brown, 2003.

Daum, Andreas W. *Kennedy in Berlin*. Washington, DC: German Historical Institute, 2008.

Dobbs, Michael. *One Minute to Midnight: Kennedy, Khrushchev, and Castro on the Brink of Nuclear War*. New York: Vintage Books, 2009.

Etzioni, Amitai. *The Hard Way to Peace: A New Strategy*. New York: Collier Books, 1962.

———. "The Kennedy Experiment." *Western Political Quarterly* 20, no. 2, part 1 (June 1967): 361–380.

Fursenko, Aleksandr, and Timothy J. Naftali. *Khrushchev's Cold War: The Inside Story of an American Adversary*. New York: W. W. Norton, 2006.

———. *"One Hell of a Gamble": Khrushchev, Castro, and Kennedy, 1958–1964.* New York: W. W. Norton, 1997.

Gaddis, John Lewis. *The Cold War: A New History.* New York: Penguin Press, 2005.

———. *Strategies of Containment: A Critical Appraisal of American National Security Policy During the Cold War.* New York: Oxford University Press, 2005.

Jervis, Robert. "Cooperation Under the Security Dilemma." *World Politics* 30, no. 2 (1978): 167–214.

———. *Perception and Misperception in International Politics.* Princeton, NJ: Princeton University Press, 1976.

Jervis, Robert, Richard Ned Lebow, and Janice Gross Stein. *Psychology and Deterrence.* Baltimore: Johns Hopkins University Press, 1985.

Kennedy, John F. *"Let the Word Go Forth": The Speeches, Statements, and Writings of John F. Kennedy.* Edited by Theodore C. Sorensen. New York: Delacorte Press, 1988.

———. *Profiles in Courage.* New York: Harper & Row, 1964.

Kennedy, John F., and Nikita Sergeevich Khrushchev. *Top Secret: The Kennedy-Khrushchev Letters.* Edited by Thomas Fensch. The Woodlands, TX: New Century Books, 2001.

Krepon, Michael, and Dan Caldwell, eds. *The Politics of Arms Control Treaty Ratification.* New York: St. Martin's Press, 1991.

Larson, Deborah Welch. *Anatomy of Mistrust: U.S.-Soviet Relations During the Cold War.* Ithaca, NY: Cornell University Press, 1997.

Leaming, Barbara. *Jack Kennedy: The Education of a Statesman.* New York: W. W. Norton, 2006.

Liddell Hart, Basil Henry. *Deterrent or Defense: A Fresh Look at the West's Military Position.* New York: Praeger, 1960.

Loeb, Benjamin S. "The Limited Test Ban Treaty." In *The Politics of Arms Control Treaty Ratification,* ed. Michael Krepon and Dan Caldwell, 167–227. New York: St. Martin's Press, 1991.

Mastny, Vojtech. "The 1963 Nuclear Test Ban Treaty: A Missed Opportunity for Détente?" *Journal of Cold War Studies* 10, no. 1 (Winter 2008): 3–25.

Nathan, James A., ed. *The Cuban Missile Crisis Revisited.* New York: St. Martin's Press, 1992.

Norris, Robert S., and Hans M. Kristensen. "Global Nuclear Weapons Inventories, 1945–2010." *Bulletin of the Atomic Scientists* 66, no. 4 (2010): 77–83.

Powaski, Ronald E. *March to Armageddon: The United States and the Nuclear Arms Race, 1939 to the Present.* New York: Oxford University Press, 1987.

Preble, Christopher A. "'Who Ever Believed in the "Missile Gap"?': John F. Kennedy and the Politics of National Security." *Presidential Studies Quarterly* 33, no. 4 (December 2003): 801–826.

Reeves, Richard. *President Kennedy: Profile of Power.* New York: Simon & Schuster, 1993.

Rhodes, Richard. *Arsenals of Folly: The Making of the Nuclear Arms Race.* New York: Alfred A. Knopf, 2007.

Richter, James G. "Perpetuating the Cold War: Domestic Sources of International Patterns of Behavior." *Political Science Quarterly* 107, no. 2 (Summer 1992): 271–301.

Rosenberg, David Alan. "The Origins of Overkill: Nuclear Weapons and American Strategy." *International Security* 7, no. 4 (Spring 1983): 3–71.

Schlesinger, Arthur M. *A Thousand Days: John F. Kennedy in the White House.* Boston: Houghton Mifflin, 1965.

Seaborg, Glenn T. *Kennedy, Khrushchev, and the Test Ban.* Edited by Benjamin S. Loeb. Berkeley: University of California Press, 1981.

Sorensen, Theodore C. *Counselor: A Life at the Edge of History.* New York: Harper, 2008.

———. *Decision-Making in the White House: The Olive Branch or the Arrows.* New York: Columbia University Press, 2005.

———. *Kennedy.* New York: Harper & Row, 1965.

———. *The Kennedy Legacy.* New York: Macmillan, 1969.

Taubman, William. *Khrushchev: The Man and His Era.* New York: W. W. Norton, 2004.

Terchek, Ronald. *The Making of the Test Ban Treaty.* The Hague: M. Nijhoff, 1970.

Thomas, Evan. *Ike's Bluff: President Eisenhower's Secret Battle to Save the World.* New York: Little, Brown, 2012.

Trachtenberg, Marc. *The Cold War and After: History, Theory, and the Logic of International Politics.* Princeton, NJ: Princeton University Press, 2012.

———. *A Constructed Peace: The Making of the European Settlement, 1945–1963.* Princeton, NJ: Princeton University Press, 1999.

———. *History and Strategy.* Princeton, NJ: Princeton University Press, 1991.

Trachtenberg, Marc, ed. *The Development of American Strategic Thought, 1945–1969.* Vol. 4. New York: Garland, 1988.

Weiner, Tim. *Legacy of Ashes: The History of the CIA.* New York: Doubleday, 2007.

Wells, Samuel F., Jr. "The Origins of Massive Retaliation." *Political Science Quarterly* 96, no. 1 (Spring 1981): 31–52.

Wenger, Andreas, and Marcel Gerber. "John F. Kennedy and the Limited Test Ban Treaty: A Case Study of Presidential Leadership." *Presidential Studies Quarterly* 29, no. 2 (June 1999): 460–487.

X. "The Sources of Soviet Conduct." *Foreign Affairs* 25, no. 4 (July 1947): 566–582.

Zubok, Vladislav, and Konstantin Pleshakov. *Inside the Kremlin's Cold War: From Stalin to Khrushchev.* Cambridge, MA: Harvard University Press, 1996.

NOTES

PREFACE

1. Graham T. Allison and Philip Zelikow, *Essence of Decision: Explaining the Cuban Missile Crisis,* 2nd ed. (New York: Longman, 1999), 240.
2. In a private letter to Khrushchev on October 28, a day after the incident, Kennedy wrote, "I regret this incident and will see to it that every precaution is taken to prevent recurrence." John F. Kennedy and Nikita Sergeevich Khrushchev, *Top Secret: The Kennedy-Khrushchev Letters,* ed. Thomas Fensch (The Woodlands, TX: New Century Books, 2001), 341.
3. Jeffrey Sachs, "The Great Convergence," *The Reith Lectures,* BBC Radio 4 (2007).

CHAPTER 1: THE QUEST FOR PEACE

1. Robert Dallek, *An Unfinished Life: John F. Kennedy, 1917–1963* (Boston: Little, Brown, 2003); Richard Reeves, *President Kennedy: Profile of Power* (New York: Simon & Schuster, 1993); Alan Brinkley, *John F. Kennedy* (New York: Times Books, 2012); Barbara Leaming, *Jack Kennedy: The Education of a Statesman* (New York: W. W. Norton, 2006).

2. Winston Churchill, *The World Crisis, 1911–1918* (New York: Free Press, 2005).

3. Martin Gilbert, *Winston Churchill* (London: Oxford University Press, 1966).

4. Neville Chamberlain, "Peace for Our Time" (speech, London, September 30, 1938), Britannia Historical Documents, http://www.britannia.com/history/docs/peacetime.html.

5. David Nasaw, *The Patriarch: The Remarkable Life and Turbulent Times of Joseph P. Kennedy* (New York: Penguin Press, 2012).

6. John F. Kennedy, *Why England Slept* (New York: W. Funk, 1940).

7. Stephen J. Majeski, "Arms Races as Iterated Prisoner's Dilemma Games," *Mathematical Social Sciences* 7, no. 3 (June 3, 1984): 253–266; Robert M. Axelrod, *The Evolution of Cooperation* (New York: Basic Books, 2006).

8. Robert Jervis, "Cooperation Under the Security Dilemma," *World Politics* 30, no. 2 (1978): 167–214.

9. Ibid., 189.

10. "Deployment by Country, 1951–1977," National Security Archive, http://www.gwu.edu/~nsarchiv/news/19991020/04-46.htm.

11. W. H. Lawrence, "Churchill Urges Patience in Coping with Red Dangers," *New York Times,* June 27, 1954, 1.

12. Winston Churchill, *Winston S. Churchill: His Complete Speeches, 1897–1963,* ed. Robert Rhodes James (New York: Chelsea House Publishers, 1974), 8257.

13. John F. Kennedy, "Speech of Senator John F. Kennedy, Civic Auditorium, Seattle, WA" (speech, Seattle, September 6, 1960), *The American Presidency Project,* ed. Gerhard Peters and John T. Woolley, http://www.presidency.ucsb.edu/ws/?pid=25654.

14. Amitai Etzioni, *The Hard Way to Peace: A New Strategy* (New York: Collier Books, 1962).

15. Ibid., 102.

16. John F. Kennedy, "Inaugural Address" (speech, Washington, DC, January 20, 1961), Miller Center, http://millercenter.org/president/speeches/detail/3365.

17. Quoted in Christopher A. Preble, "'Who Ever Believed in the "Missile Gap"?': John F. Kennedy and the Politics of National Security," *Presidential Studies Quarterly* 33, no. 4 (December 2003): 805–806.

18. Kennedy and Khrushchev, *Top Secret,* 70.

19. Robert S. Norris and Hans M. Kristensen, "Global Nuclear Stockpiles, 1945–2006," *Bulletin of the Atomic Scientists* 62, no. 4 (2006): 66.

20. John Lewis Gaddis, *Strategies of Containment: A Critical Appraisal of American National Security Policy During the Cold War* (New York: Oxford University Press, 2005), 198–236; Lawrence Freedman, *The Evolution of Nuclear Strategy* (New York: St. Martin's Press, 1989); Joseph M. Siracusa and David G. Coleman, "Scaling the Nuclear Ladder: Deterrence from Truman to Clinton," *Australian Journal of International Affairs* 54, no. 3 (2000): 285–288.

21. David Alan Rosenberg, "The Origins of Overkill: Nuclear Weapons and American Strategy," *International Security* 7, no. 4 (Spring 1983): 66.

22. Marifeli Pérez-Stable, *The Cuban Revolution: Origins, Course and Legacy* (New York: Oxford University Press, 2012).

23. Aleksandr Fursenko and Timothy J. Naftali, *"One Hell of a Gamble": Khrushchev, Castro, and Kennedy, 1958–1964* (New York: W. W. Norton, 1997), 82–100; Michael R. Beschloss, *The Crisis Years: Kennedy and Khrushchev, 1960–1963* (New York: Edward Burlingame Books, 1991); Trumbull Higgins, *The Perfect Failure: Kennedy, Eisenhower, and the CIA at the Bay of Pigs* (New York: W. W. Norton, 1987); Piero Gleijeses, "Ships in the Night: The CIA, the White House and the Bay of Pigs," *Journal of Latin American Studies* 27, no. 1 (February 1995): 1–42.

24. Beschloss, *The Crisis Years,* 145.

25. Dallek, *An Unfinished Life,* 368.

26. John F. Kennedy, "Special Message to the Congress on Urgent National Needs" (speech, Washington, DC, May 25, 1961), *The American Presidency Project,* ed. Gerhard Peters and John T. Woolley, http://www.presidency.ucsb.edu/ws/?pid=8151.

27. Fursenko and Naftali, *"One Hell of a Gamble,"* 146–150, 156–158; Edward Lansdale, "Operation Mongoose: The Cuba Project," February 20, 1962, Cuban History Archive, http://www.marxists.org/history/cuba/subject/cia/mongoose/c-project.htm.

28. Kennedy and Khrushchev, *Top Secret,* 2.

29. Ibid., 2–3.

30. Ibid., 3.

31. Ibid., 13–14.

32. Ibid., 16.

33. Ibid., 18–19.

34. Ibid., 25.

35. *Foreign Relations of the United States: Diplomatic Papers; The Conference of Berlin (Potsdam Conference, 1945),* 2 vols. (Washington, DC: U.S. Government Printing Office, 1960).

36. Fraser J. Harbutt, *Yalta 1945: Europe and America at the Crossroads* (New York: Cambridge University Press, 2010).

37. Frederick H. Gareau, "Morgenthau's Plan for Industrial Disarmament in Germany," *Western Political Quarterly* 14, no. 2 (June 1961): 517–534.

38. James F. Byrnes, "Restatement of Policy on Germany" (speech, Stuttgart, September 6, 1946), United States Diplomatic Mission to Germany, http://usa.usembassy.de/etexts/ga4-460906.htm.

39. Marc Trachtenberg, *A Constructed Peace: The Making of the European Settlement, 1945–1963* (Princeton, NJ: Princeton University Press, 1999), 146–149.

40. Ibid.

41. Ibid., 212–213.

42. Quoted in Fursenko and Naftali, *"One Hell of a Gamble,"* 123.
43. Quoted in Beschloss, *The Crisis Years,* 224.
44. Quoted in ibid., 217.
45. Kennedy and Khrushchev, *Top Secret,* 218.
46. Quoted in Beschloss, *The Crisis Years,* 218.
47. Frederick Kempe, *Berlin 1961: Kennedy, Khrushchev, and the Most Dangerous Place on Earth* (New York: G. P. Putnam's Sons, 2011), 257.
48. Deborah Welch Larson, *Anatomy of Mistrust: U.S.-Soviet Relations During the Cold War* (Ithaca, NY: Cornell University Press, 1997), 133.
49. "Facts About the Berlin Wall," Agence France-Presse, accessed February 14, 2013, http://www.abs-cbnnews.com/features/11/09/09/facts-about-berlin-wall.
50. Quoted in Beschloss, *The Crisis Years,* 225.

CHAPTER 2: TO THE BRINK

1. Beschloss, *The Crisis Years,* 291.
2. Ibid.
3. "30 October 1961—The Tsar Bomba," Comprehensive Nuclear-Test Ban Treaty Organization Preparatory Commission, accessed February 14, 2013, http://www.ctbto.org/specials/infamous-anniversaries/30-october-1961-the-tsar-bomba/.
4. Quoted in Beschloss, *The Crisis Years,* 307.
5. Quoted in Dallek, *An Unfinished Life,* 463.
6. Roswell Gilpatric, "Address by Deputy Secretary of Defense Gilpatric to the Business Council" (1961), in Meena Bose, *Shaping and Signaling Presidential Policy: The National Security Decision Making of Eisenhower and Kennedy* (College Station: Texas A&M University Press, 1998), 149–156.
7. Allison and Zelikow, *Essence of Decision.*
8. Beschloss, *The Crisis Years,* 387.
9. Ibid., 388–389.
10. For more detailed analysis and information on the Cuban Missile Crisis, see Allison and Zelikow, *Essence of Decision*; James A. Nathan, ed., *The Cuban Missile Crisis Revisited* (New York: St. Martin's Press, 1992); Fursenko and Naftali, *"One Hell of a Gamble"*; Ernest R. May and Philip Zelikow, eds., *The Kennedy Tapes: Inside the White House During the Cuban Missile Crisis* (New York: W. W. Norton, 2002); James G. Blight, Bruce J. Allyn, David A. Welch, and David Lewis, eds., *Cuba on the Brink: Castro, the Missile Crisis, and the Soviet Collapse* (New York: Pantheon Books, 1993); Michael Dobbs, *One Minute to Midnight: Kennedy, Khrushchev, and Castro on the Brink of Nuclear War* (New York: Vintage Books, 2009).
11. John F. Kennedy, "The President's News Conference," September 13, 1962, *The American Presidency Project,* ed. Gerhard Peters and John T. Woolley, http://www.presidency.ucsb.edu/ws/?pid=8867.

12. Theodore C. Sorensen, *Kennedy* (New York: Harper & Row, 1965), 705.

13. May and Zelikow, eds., *The Kennedy Tapes,* 62.

14. Quoted in Richard Ned Lebow, "The Cuban Missile Crisis: Reading the Lessons Correctly," *Political Science Quarterly* 98, no. 3 (Autumn 1983): 443.

15. Kennedy and Khrushchev, *Top Secret,* 343–344.

16. Norman Cousins, *The Improbable Triumvirate: John F. Kennedy, Pope John, Nikita Khrushchev* (New York: W. W. Norton, 1972), 37.

17. Beschloss, *The Crisis Years,* 556.

18. Quoted in Allison and Zelikow, *Essence of Decision,* 355.

19. Quoted in Thurston Clarke, *JFK's Last Hundred Days* (New York: Penguin Press, 2013).

20. Quoted in Cousins, *The Improbable Triumvirate,* 46.

21. James A. Nathan, "The Heyday of the New Strategy: The Cuban Missile Crisis and the Confirmation of Coercive Diplomacy," in *The Cuban Missile Crisis Revisited,* ed. Nathan, 24.

22. Beschloss, *The Crisis Years,* 570.

23. Ibid., 572.

24. Kennedy and Khrushchev, *Top Secret,* 314.

25. Ibid., 348–349.

CHAPTER 3: PRELUDE TO PEACE

1. Barbara W. Tuchman, *The Guns of August* (New York: Macmillan, 1962).

2. Reeves, *President Kennedy,* 306.

3. Basil Henry Liddell Hart, *Deterrent or Defense: A Fresh Look at the West's Military Position* (New York: Praeger, 1960), 254, 257.

4. Sharon Bertsch McGrayne, *The Theory That Would Not Die: How Bayes' Rule Cracked the Enigma Code, Hunted Down Russian Submarines, & Emerged Triumphant from Two Centuries of Controversy* (New Haven, CT: Yale University Press, 2011).

5. Kennedy and Khrushchev, *Top Secret,* 487.

6. Cousins, *The Improbable Triumvirate,* 113.

7. Ibid., 114.

8. Ibid., 55.

9. James G. Richter, "Perpetuating the Cold War: Domestic Sources of International Patterns of Behavior," *Political Science Quarterly* 107, no. 2 (Summer 1992): 273–274.

10. Kennedy and Khrushchev, *Top Secret,* 422.

CHAPTER 4: THE RHETORIC OF PEACE

1. John F. Kennedy, *Profiles in Courage* (New York: Harper & Row, 1964).

2. Winston Churchill, "The Few" (speech, London, August 20, 1940), the

Churchill Centre and Museum, http://www.winstonchurchill.org/learn
/speeches/speeches-of-winston-churchill/1940-finest-hour/113-the-few.

3. John F. Kennedy, "Proclamation 3525—Declaring Sir Winston Churchill an
Honorary Citizen of the United States of America," April 9, 1963, *The American Presidency Project,* ed. Gerhard Peters and John T. Woolley, http://www
.presidency.ucsb.edu/ws/?pid=24064.

4. Reeves, *President Kennedy,* 41.

5. McGeorge Bundy, recorded interview by Richard Neustadt, March 1964,
99–100, John F. Kennedy Library Oral History Program.

6. Theodore C. Sorensen, *Counselor: A Life at the Edge of History* (New York:
Harper, 2008), 27.

7. John XXIII, *Pacem in Terris,* Encyclical of Pope John XXIII on Establishing Universal Peace in Truth, Justice, Charity, and Liberty, April 11, 1963,
Papal Encyclicals Online, http://www.papalencyclicals.net/John23/j23pacem
.htm, sec. 126.

8. Winston Churchill, "The Sinews of Peace" (speech, Fulton, Missouri,
March 5, 1946), *Britannia Historical Documents,* http://www.britannia.com
/history/docs/sinews1.html.

9. Dwight D. Eisenhower, "Chance for Peace" (speech, Washington, DC,
April 16, 1953), Miller Center, http://millercenter.org/president/speeches
/detail/3357.

10. Evan Thomas, *Ike's Bluff: President Eisenhower's Secret Battle to Save the
World* (New York: Little, Brown, 2012), 65.

11. Dwight D. Eisenhower, "Atoms for Peace" (speech, New York, December 8, 1953), Miller Center, http://millercenter.org/president/speeches
/detail/3358.

12. Winston Churchill, Speech to the House of Commons, May 11, 1945,
Parliamentary Debates, Commons, vol. 515 cc883-1004, http://hansard
.millbanksystems.com/commons/1953/may/11/foreign-affairs#S5CV0515P0
_19530511_HOC_220.

13. Quoted in Leaming, *Jack Kennedy,* 213.

14. Dwight D. Eisenhower, "Farewell Address" (speech, Washington, DC, January 17, 1961), Miller Center, http://millercenter.org/president/speeches
/detail/3361.

15. John F. Kennedy, "Inaugural Address" (speech, Washington, DC, January 20, 1961), Miller Center, http://millercenter.org/president/speeches
/detail/3365.

16. Winston Churchill, Speech to the House of Commons, November 3, 1953.
Parliamentary Debates, Commons, vol. 520 cc7-136, http://hansard
.millbanksystems.com/commons/1953/nov/03/debate-on-the-address-first
-day.

17. John F. Kennedy, "Address to the UN General Assembly" (speech, New
York, September 25, 1961), Miller Center, http://millercenter.org/president
/speeches/detail/5741.

18. Cousins, *The Improbable Triumvirate,* 65.

19. Ibid., 63.
20. John XXIII, *Pacem in Terris,* sec. 31.
21. Ibid., sec. 113.
22. Ibid., sec. 118.
23. Ibid., sec. 127–129.

CHAPTER 5: THE PEACE SPEECH

1. Theodore C. Sorensen, recorded interview by Carl Kaysen, April 15, 1964, 70–71, John F. Kennedy Library Oral History Program.
2. Cousins, *The Improbable Triumvirate,* 116.
3. Sorensen, recorded interview by Kaysen, 72.
4. William C. Foster, recorded interview by Charles T. Morrissey, August 5, 1964, 32, John F. Kennedy Library Oral History Program.
5. John F. Kennedy, "American University Commencement" (speech, Washington, DC, June 10, 1963), Miller Center, http://millercenter.org/president/speeches/detail/3374.
6. Churchill, Speech to the House of Commons, November 3, 1953.
7. John F. Kennedy, "Address on Civil Rights" (speech, Washington, DC, June 11, 1963), Miller Center, http://millercenter.org/president/speeches/detail/3375.

CHAPTER 6: THE CAMPAIGN FOR PEACE

1. "A Strategy of Peace," *Washington Post,* June 11, 1963, 14.
2. "New Hope for a Test Ban," *New York Times,* June 11, 1963, 36.
3. Walter Lippmann, "Let Live or Don't Live," *Boston Globe,* June 13, 1963, 18.
4. Roscoe Drummond, "A New Look at Cold War," *Christian Science Monitor,* June 26, 1963, 9.
5. Thomas Sorensen to McGeorge Bundy, "West European Reaction to Peace Speech," June 12, 1963, United States Information Agency, Sorensen Papers, Box 72, American University Commencement 6/10/63, John F. Kennedy Library.
6. Quoted in "British Welcome Speech," *New York Times,* June 12, 1963, 4.
7. "The President's Lead," *The Times,* June 11, 1963, 13.
8. Richard H. Crossman, "Philosophy of Peace," *The Guardian,* June 14, 1963, 20.
9. Amitai Etzioni, "The Kennedy Experiment," *Western Political Quarterly* 20, no. 2, part 1 (June 1967): 366.
10. Foreign Broadcast Information Service, World Reaction Series, "Foreign Radio and Press Reaction to the President's Foreign Policy Speech at American University on 10 June 1963," June 14, 1963, NSF/305A/President's Speeches: American University Speech, 6/10/63, 6/14/63–6/15/63, John F. Kennedy Library.

11. Arthur M. Schlesinger, *A Thousand Days: John F. Kennedy in the White House* (Boston: Houghton Mifflin, 1965), 904.

12. Ibid.

13. CIA Information Report, "Soviet Reaction to 10 June Speech of President Kennedy," June 11, 1963, NSF/305A/President's Speeches: American University Speech, 6/10/63, 6/13/63, John F. Kennedy Library.

14. "Text of Khrushchev on Kennedy Speech," Moscow TASS in English to Europe 2205, June 14, 1963, NSF/305A/President's Speeches: American University Speech, 6/10/63, 6/14/63–6/15/63, John F. Kennedy Library.

15. "Proposal Concerning the General Line of the International Communist Movement: The Letter of the Central Committee of the Communist Party of China in Reply to the Central Committee of the Communist Party of the Soviet Union of March 30, 1963," Sino-Soviet Split Document Archive, http://www.marxists.org/history/international/comintern/sino-soviet-split/cpc/proposal.htm.

16. John Milton Cooper Jr., *Woodrow Wilson* (New York: Random House, 2009), 462.

17. Schlesinger, *A Thousand Days*, 884.

18. John F. Kennedy, "'Ich bin ein Berliner' Speech" (speech, Berlin, June 26, 1963), Miller Center, http://millercenter.org/president/speeches/detail/3376.

19. Andreas W. Daum, *Kennedy in Berlin* (Washington, DC: German Historical Institute, 2008), 28, 120.

20. Theodore Windt, *Presidents and Protesters: Political Rhetoric in the 1960s* (Tuscaloosa: University of Alabama Press, 1990), 72.

21. Bundy, recorded interview by Neustadt, 29.

22. John F. Kennedy, "Address at the Free University of Berlin" (speech, Berlin, June 26, 1963), *The American Presidency Project*, ed. Gerhard Peters and John T. Woolley, http://www.presidency.ucsb.edu/ws/?pid=9310.

23. John F. Kennedy, "Address Before the Irish Parliament in Dublin" (speech, Dublin, June 28, 1963), *The American Presidency Project*, ed. Gerhard Peters and John T. Woolley, http://www.presidency.ucsb.edu/ws/?pid=9317.

24. George Bernard Shaw, *Back to Methuselah* (Project Gutenberg eBook, 2004), http://www.gutenberg.org/files/13084/13084-8.txt.

25. Edward M. Kennedy, "Address at the Public Memorial Service for Robert F. Kennedy" (speech, New York, June 8, 1969), *American Rhetoric*, http://www.americanrhetoric.com/speeches/ekennedytributetorfk.html.

26. John F. Kennedy, "Remarks in Naples at NATO Headquarters" (speech, Naples, July 2, 1963), *The American Presidency Project*, ed. Gerhard Peters and John T. Woolley, http://www.presidency.ucsb.edu/ws/?pid=9332.

27. Sorensen, *Kennedy*, 733.

28. Glenn T. Seaborg, *Kennedy, Khrushchev, and the Test Ban*, ed. Benjamin S. Loeb (Berkeley: University of California Press, 1981), 179.

29. Cousins, *The Improbable Triumvirate*, 97.

30. Seaborg, *Kennedy, Khrushchev, and the Test Ban,* 227.

31. Quoted in ibid.

32. NSC Action 2468, Instructions for Harriman Mission, July 9, 1963, NSF/265 /ACDA: Disarmament, Subjects, Nuclear Test Ban Treaty, Harriman Trip to Moscow, Part B, John F. Kennedy Library.

33. Seaborg, *Kennedy, Khrushchev, and the Test Ban,* 237.

34. Andreas Wenger and Marcel Gerber, "John F. Kennedy and the Limited Test Ban Treaty: A Case Study of Presidential Leadership," *Presidential Studies Quarterly* 29, no. 2 (June 1999), 478.

35. Treaty Banning Nuclear Weapon Tests in the Atmosphere, in Outer Space and Under Water, U.S.-U.K.-U.S.S.R., Aug. 5, 1963, 14 U.S.T. 1313.

36. Foster, recorded interview by Morrissey, 30.

37. Adrian S. Fisher, recorded interview by Frank Sieverts, May 13, 1964, 77, John F. Kennedy Library Oral History Program.

CHAPTER 7: CONFIRMING THE TREATY

1. Cousins, *The Improbable Triumvirate,* 128.

2. Seaborg, *Kennedy, Khrushchev, and the Test Ban,* 264.

3. Cousins, *The Improbable Triumvirate,* 135.

4. John F. Kennedy, "Address on the Nuclear Test Ban Treaty" (speech, Washington, DC, July 26, 1963), Miller Center, http://millercenter.org/president /speeches/detail/3377.

5. Dallek, *An Unfinished Life,* 622.

6. Ibid., 345.

7. Wenger and Gerber, "John F. Kennedy and the Limited Test Ban Treaty," 478.

8. Benjamin S. Loeb, "The Limited Test Ban Treaty," in *The Politics of Arms Control Treaty Ratification,* ed. Michael Krepon and Dan Caldwell (New York: St. Martin's Press, 1991), 188.

9. Maxwell Taylor to Dean Rusk, July 27, 1963, NSF/264/ACDA: Disarmament, Subjects, Nuclear Test Ban Treaty, Congressional Relations, 5/63–7/63, John F. Kennedy Library.

10. Seaborg, *Kennedy, Khrushchev, and the Test Ban,* 265.

11. Loeb, "The Limited Test Ban Treaty," in *The Politics of Arms Control Treaty Ratification,* 189.

12. Nuclear Test Ban Treaty: Hearings Before the Committee on Foreign Relations, United States Senate, Eighty-eighth Congress, First Session, on The Treaty Banning Nuclear Weapon Tests in the Atmosphere, In Outer Space, and Underwater, Signed at Moscow on August 5, 1963, on Behalf of the United States of America, the United Kingdom of Great Britain and Northern Ireland, and the Union of Soviet Socialist Republics, 88th Cong. 276 (1963) (statement by General Maxwell Taylor, Chairman of the Joint Chiefs of Staff).

13. "Text of Scientists Statement Supporting Test Ban Treaty," POF/100/JFKL, John F. Kennedy Library.

14. Seaborg, *Kennedy, Khrushchev, and the Test Ban*, 273.

15. Nuclear Test Ban Treaty: Hearings Before the Committee on Foreign Relations, United States Senate, Eighty-eighth Congress, First Session, 88th Cong. 846–848 (1963) (letter from Dwight D. Eisenhower, Former President of the United States).

16. Sorensen, recorded interview by Kaysen, 83.

17. John F. Kennedy, "Letter to Senate Leaders Restating the Administration's Views on the Nuclear Test Ban Treaty," September 11, 1963, *The American Presidency Project,* ed. Gerhard Peters and John T. Woolley, http://www .presidency.ucsb.edu/ws/?pid=9403.

18. Seaborg, *Kennedy, Khrushchev, and the Test Ban*, 280.

19. Loeb, "The Limited Test Ban Treaty," in *The Politics of Arms Control Treaty Ratification,* 205.

20. John F. Kennedy, "Statement by the President Following the Senate Vote on the Nuclear Test Ban Treaty," September 24, 1963, *The American Presidency Project,* ed. Gerhard Peters and John T. Woolley, http://www.presidency .ucsb.edu/ws/?pid=9426.

21. To John F. Kennedy, "Summary of Governmental Reactions to Test Ban Treaty (excludes United States, UK, and USSR)," NSF/264/ACDA: Disarmament, Subjects, Nuclear Test Ban Treaty, Foreign Reaction, 7/63–10/63, John F. Kennedy Library.

22. John F. Kennedy, "Address to the UN General Assembly" (speech, New York, September 20, 1963), Miller Center, http://millercenter.org/president /speeches/detail/5764.

23. Vojtech Mastny, "The 1963 Nuclear Test Ban Treaty: A Missed Opportunity for Détente?" *Journal of Cold War Studies* 10, no. 1 (Winter 2008): 8.

24. Etzioni, "The Kennedy Experiment," 367.

25. Kennedy and Khrushchev, *Top Secret,* 562.

26. Ibid., 566.

27. Quoted in William Taubman, *Khrushchev: The Man and His Era* (New York: W. W. Norton, 2004), 13.

CHAPTER 8: THE HISTORIC
MEANING OF KENNEDY'S PEACE INITIATIVE

1. "Doomsday Clock," *Bulletin of the Atomic Scientists,* accessed February 11, 2013, http://www.thebulletin.org/content/doomsday-clock/timeline.

2. Treaty on the Non-Proliferation of Nuclear Weapons, U.S-U.K.-U.S.S.R., July 1, 1968, 21 U.S.T. 483, 729 U.N.T.S. 161.

3. John F. Kennedy, "The President's News Conference," March 21, 1963, *The American Presidency Project,* ed. Gerhard Peters and John T. Woolley, http://www.presidency.ucsb.edu/ws/?pid=9124.

4. Thomas Graham Jr., "Avoiding the Tipping Point," review of *The Nuclear*

Tipping Point: Why States Reconsider Their Nuclear Choices, ed. Kurt M. Campbell, Robert J. Einhorn, and Mitchell B. Reiss (Washington, DC: Brookings Institution Press, 2004).

5. See in particular the Nuclear Security Project, http://www.nuclearsecurity project.org/, and the call by Henry Kissinger, Sam Nunn, George Shultz, and William Perry for a nuclear-free world.

6. For a brief overview of the period after 1963, see Raymond L. Garthoff, *Détente and Confrontation: American-Soviet Relations from Nixon to Reagan* (Washington, DC: Brookings Institution, 1994); Gaddis, *Strategies of Containment*; Shane J. Maddock, *Nuclear Apartheid: The Quest for American Atomic Supremacy from World War II to the Present* (Chapel Hill: University of North Carolina Press, 2010); Richard W. Stevenson, *The Rise and Fall of Détente: Relaxations of Tension in US-Soviet Relations, 1953–84* (Urbana: University of Illinois Press, 1985).

7. "Strategic Arms Limitation Talks (SALT I)," Arms Control Association, accessed February 19, 2013, http://www.armscontrol.org/documents/salt.

8. "Strategic Arms Limitation Talks II (SALT II)," Arms Control Association, accessed February 19, 2013, http://www.armscontrol.org/documents /salt2.

9. "Doomsday Clock," *Bulletin of the Atomic Scientists.*

10. Robert S. Norris and Hans M. Kristensen, "Global Nuclear Weapons Inventories, 1945–2010," *Bulletin of the Atomic Scientists* 66, no. 4 (2010): 81–82; Robert S. Norris and Hans M. Kristensen, "Russian Nuclear Forces, 2010," *Bulletin of the Atomic Scientists* 66, no. 1 (2010): 74–81; Hans M. Kristensen and Robert S. Norris, "Russian Nuclear Forces, 2011," *Bulletin of the Atomic Scientists* 67, no. 3 (2011): 67–74; Hans M. Kristensen and Robert S. Norris, "Russian Nuclear Forces, 2012," *Bulletin of the Atomic Scientists* 68, no. 2 (2012): 87–97; Hans M. Kristensen and Robert S. Norris, "US Nuclear Forces, 2011," *Bulletin of the Atomic Scientists* 67, no. 2 (2011): 66–76; Hans M. Kristensen and Robert S. Norris, "US Nuclear Forces, 2012," *Bulletin of the Atomic Scientists* 68, no. 3 (2012): 84–91.

11. Norris and Kristensen, "Global Nuclear Stockpiles, 1945–2006," 66.

12. U.S. Department of Energy, Nevada Operations Office, *United States Nuclear Tests, July 1945 through September 1992* (Las Vegas: U.S. Department of Energy, Nevada Operations Office, 2000).

13. Norris and Kristensen, "Global Nuclear Weapons Inventories, 1945–2010," 81–82; Thomas B. Cochran, William M. Arkin, Robert S. Norris, and Jeffrey I. Sands, *Nuclear Weapons Databook,* vol. 4, *Soviet Nuclear Weapons* (New York: Harper & Row, 1989); "Table of USSR/Russian ICBM Forces," Natural Resources Defense Council, accessed March 22, 2013, http://www .nrdc.org/nuclear/nudb/datab4.asp.

14. Quoted in Murrey Marder, "Summit Clouded by Watergate," *New York Times,* July 4, 1974.

15. "START I at a Glance," Arms Control Association, accessed February 19, 2013, http://www.armscontrol.org/factsheets/start1.

16. "Strategic Arms Reduction Treaty II (START II)," Arms Control Association, accessed February 19, 2013, http://www.armscontrol.org/node/2494.

17. "New START," U.S. State Department, accessed February 19, 2013, http://www.state.gov/t/avc/newstart/index.htm.

18. *Encyclopædia Britannica Online*, s.v. "Vietnam War," accessed February 19, 2013, http://www.britannica.com/EBchecked/topic/628478/Vietnam-War.

19. *Encyclopædia Britannica Online*, s.v. "Soviet Invasion of Afghanistan," accessed February 19, 2013, http://www.britannica.com/EBchecked/topic/1499983/Soviet-invasion-of-Afghanistan.

20. The most prescient accounts were by Soviet dissident Andreii Amalrik in *Will the Soviet Union Survive Until 1984?* in 1970, and by French sociologist Emmanuel Todd in *The Final Fall: An Essay on the Decomposition of the Soviet Sphere* in 1976 (Andrei Almarik, *Will the Soviet Union Survive Until 1984?* [New York: Harper & Row, 1970]; Emmanuel Todd, *La Chute Finale: Essai sur la Décomposition de la Sphère Soviétique* [Paris: R. Laffont, 1976]).

21. John Maynard Keynes, *Essays in Persuasion* (London: Macmillan, 1931), 306.

22. X [George Kennan], "The Sources of Soviet Conduct," *Foreign Affairs* 25, no. 4 (July 1947), 582. Emphasis added.

23. Ibid.

CHAPTER 9: LET US TAKE OUR STAND

1. John F. Kennedy, "Address at Rice University in Houston on the Nation's Space Effort" (speech, Houston, September 12, 1962), *The American Presidency Project,* ed. Gerhard Peters and John T. Woolley, http://www.presidency.ucsb.edu/ws/?pid=8862.

2. John F. Kennedy, "Address on Civil Rights," Miller Center.

3. "Doomsday Clock," *Bulletin of the Atomic Scientists.*

4. George C. Marshall, "Commencement Address at Harvard University (the Marshall Plan Speech)" (speech, Cambridge, June 5, 1947), George C. Marshall Foundation, http://www.marshallfoundation.org/library/MarshallPlanSpeechfromRecordedAddress_000.html.

5. Organisation for Economic Co-operation and Development, Development Assistance Committee.

6. Barack Obama, "Remarks by the President on a New Beginning" (speech, Cairo, June 4, 2009), White House Office of the Press Secretary, http://www.whitehouse.gov/the-press-office/remarks-president-cairo-university-6-04-09.

INDEX

Page numbers in *italics* refer to illustrations.

Acheson, Dean, 30*n*, 79

Adams, Sherman, 59

Adenauer, Konrad, 21, 37, 47, 97, 134, 142

Afghanistan, 164, 165
  coups and countercoups in, 150
  Soviet invasion of, 143, 150, 151
  U.S. military action in, 163

Africa, 164, 165, 167, 203

Air Force, U.S., xi, 8

Alabama, University of, integration of, 86, 87, 88

Alabama National Guard, 86*n*, 88

Alliance for Progress, 164

Allies (postwar), 133, 174, 205
  JFK's peace campaign among, 90, 91, 96–101, 119
  U.S. military bases and missile site placement, 14, 28, 29, 88, 95

Allies (World War II), 17, 19–20, 80, 82, 171
  Eisenhower as supreme commander of, 57

Amalrik, Andreii, 230*n*

American Revolution, 182–83, 186

American University, xiv, 71–74, 86–87, 91, 94, 97*n*, 100, 103, 128, 132, 170

American University speech (Kennedy), *see* Peace Speech of 1963

Anderson, George W., Jr., *120*

Annan, Kofi, xiv

Antarctic region, 204

Anti-Ballistic Missile (ABM)
     Treaty, 143
anti-ballistic missiles (ABMs),
     143, 145, 146
anti-communism, 8, 100, 151, 155
Archimedes, 133, 210
Arms Control and Disarmament
     Agency, 72, 111
Army, U.S., 8, 42–43, 182
Asia, 161, 164, 165, 167, 198
Atomic Energy Commission, 72,
     121
"Atoms for Peace" speech
     (Eisenhower), 53, 60
Australia, 181
Austrian peace treaty, 58

Back to Methuselah (Shaw), 104
Ball, George, 30n, 31
ballistic missiles, 12, 32, 143, 146,
     204
Baltic Sea, 54
Barry, John, 183
Baruch, Bernard, 191
Basic Principles of Relations
     Between United States and
     the USSR, 143
Batista, Fulgencio, 14
Bay of Pigs invasion (1961), xiii,
     13–18, 27, 65
     Cuban expatriates involved in,
         15, 18
     failure of, 15–18, 21, 28, 40, 42
     JFK's denial of involvement in,
         15–18
     lack of U.S. air support in, 15, 32
BBC, xiv, 94
Beria, Lavrentiy, 60

Berlin, 192, 194, 198, 203
     bombing of, 100
     East/West division of, 19,
         22–25, 99
     1948–49 U.S. airlift in, 57
     Soviet demands over, 19, 22–25,
         27, 28, 32, 33, 47
     see also East Berlin; West
         Berlin
Berlin, Free University of, 100
Berlin Speech (Kennedy), 97–101,
     98, 157
Berlin Wall, 24, 27, 40, 97, 99, 187
     collapse of, 151
Bermuda summit, 60–61
Beschloss, Michael, 16
Birmingham, Ala., civil rights
     demonstrations in, 87
Bismarck, Otto von, 8
Bissell, Richard, 40
Bohlen, Charles, 40
Bolshevik Revolution, 163
Bose, Meena, 60n
Bradbury, Norris, 123
Bradford, William, 159
Bradlee, Ben, 16
Brezhnev, Leonid, 135
     Nixon's relations with, 142–43
Britain, Battle of, 51
Brown v. Board of Education, 86n
Brundtland Commission of
     1987, 162
Bulletin of the Atomic Scientists,
     137, 144–45, 160
     see also Doomsday Clock
Bundy, McGeorge, 26, 30n, 31,
     34n, 39, 52, 72
Byrd, Robert, 74, 170

Canada, 166, 176

capitalism, 41, 153, 154, 173

Carter, Jimmy, 143–44

Castro, Fidel, 14–16, 35
    attempted assassination of, 16
    successful Cuban insurgency
        of, 14–15

Central Intelligence Agency
    (CIA), 29, 62, 95, 121
    Cuban invasion plan of, 13–18, 40
    mujahideen funded by, 150
    resignations following Bay of
        Pigs in, 40
    secret operations of, 150
    stepped-up operations of, 144

Chamberlain, Neville, 55–56
    appeasement by, 5–6, 9, 71, 74

"Chance for Peace" speech
    (Eisenhower), 53, 56–60, 63

Chernobyl nuclear disaster, 152

China, People's Republic of, 35
    economic relations between
        U.S. and, 143
    first nuclear test of, 110
    Nixon's visit to, 142–43
    nuclear aspirations and
        acquisition of weapons, 37,
        85, 93, 107, 127, 140, 142
    reforms of, 148
    Soviet relations with, 93, 95,
        107, 108, 127, 142, 194
    U.S. relations with, 142–43
    victory over Nationalist China
        by, 57

Chinese Communism, 57, 194

Christianity, 101, 164, 186

Christian Science Monitor, 93

Churchill, Winston, 4–6, 8, 104, 151

Eisenhower and, 60–61
    JFK influenced by, 4–5, 6, 8–10,
        11, 41, 42, 45, 51, 54, 56, 63, 64,
        66, 74, 80, 82
    negotiation through strength
        urged by, 9–10, 11, 45, 56, 63
    peace initiative of, 54–56, 61
    as prime minister, 4–5
    realism of, 51, 54, 66, 83
    rhetorical power of, 51, 53, 54
    writing of, 4, 42
    see also "Sinews of Peace"
        speech

Citizens' Committee for a Nuclear
    Test Ban, 113

Civil Rights Act of 1964, 89

civil rights movement, 86–87
    demonstrations in, 87
    integration of schools in,
        86–87, 88
    JFK's support of, 86–90, 207–8
    Supreme Court rulings in, 86n
    see also integration; racism

Civil Rights Speech of 1963
    (Kennedy), xiii–xiv, 70,
        88–90, 132

climate change, 161

Cold War, xv–xvi, 3–4, 41, 45, 54,
    67, 76, 84, 92, 96, 97, 129, 136,
    138, 148, 160, 172
    absence of trust in, 6–7, 10–11,
        18–19, 23, 26–29, 44
    conflict points in, xi–xii, 6,
        10, 13–21, 24–25, 26–38,
        56–59, 61
    ending of, 138, 139, 151–54, 168
    global relations still shaped
        by, xii

Cold War (*cont'd*)
  nuclear war avoided in, xi–xiii,
    xvi, 35–38
  ongoing dynamics of, 137, 144,
    149
  "proxy wars" resulting from,
    xii, 40, 150, 168
  psychological mindset of, 3,
    10–11, 14, 61
  Reagan intensification of, 144
  shared values on both sides in,
    70–71, 81–83, 174–75
  threat of war in, 3, 6–10, 13–14,
    22–23, 29–35, 58, 168–69
  turning point in, xii–xiii, 25, 40
colonialism, 59, 118, 202
Columbia University, xiv
communism, 79, 99–100, 153, 155,
    169, 174, 187
  capitalism vs., 41, 153, 154
  Democrats called "soft" on,
    10, 12
  in Eastern Europe, 20, 59, 92,
    175
  economic doctrine of, 93, 177
  failure of, xii, 155
  global, xiii, 153, 177, 198
  in Latin America, 35
  see also Chinese Communism;
    Soviet Communist Party
Conference of Mayors, U.S., 72
Congo, 176, 187, 188, 200, 201, 209
Congress, U.S., 62, 79, 91, 108,
    133–34, 165, 183
  see also Senate, U.S.
Constitution, U.S., 89, 117, 198
Constructed Peace, A
    (Trachtenberg), 21

Cork, *102*, 187
Cousins, Norman, 112–13
  in campaign for nuclear treaty
    support, 113–14
  as JFK emissary, 46–47, 68,
    71–72, 107, 113–14
Crossman, Richard, 94
Cuba, xii, 192, 198
  CIA planned invasion of, 13–18
  placement of Soviet missiles
    in, 14
  Soviet alliance with, 14–18, 27,
    28–38, 134
  successful Castro insurgency
    in, 14–15, 203
  U.S. assets seized by, 14
  U.S. trade embargo of, 15
  see also Bay of Pigs invasion;
    Cuban Missile Crisis
Cuban Missile Crisis (1962), xi,
    xiii, 29–38, *31*, 42, 53, 108, 204
  causes of, 16, 29–32
  escalation of, 29–33
  JFK's actions in, 29–38, 43–44,
    46, 48, 52, 121
  Khrushchev's actions in, 14, 16,
    29–38, 46, 48, 135
  lessons of, 83–84
  naval quarantine of Soviet ships
    in, 33, 121
  Soviet missiles installed by
    stealth in, 14, 29–33, 121
  threat of global war avoided in,
    29–38, 40, 48, 65, 67
  U.S. no-invasion pledge in, 33, 35
  withdrawal of Soviet missiles
    in, 33–35, 107, 146
Cuban Revolution, 14–15, 203

Czechoslovakia, 5, 20
  Prague Spring, 142

Daily Mail (London), 94
Dallek, Robert, 121*n*
Daum, Andreas W., 97*n*
Dean, Arthur H., 106*n*
deforestation, 162
de Gaulle, Charles, 46
democracy, 45, 51
  military as threat to, 62
Democratic National Convention
  of 1956, 134*n*
Democratic Party, 10, 12, 35, 119
  Southern wing of, 119
Depression, Great, 160
Dillon, Douglas, 30*n*, *31*, 39
Dirksen, Everett, 125, 134
Dobrynin, Anatoly, 34
Doomsday Clock, 137–38, *138*, 144,
  160
drones, 164
drought, 164
Dr. Strangelove (film), 121*n*
Drummond, Roscoe, 93
Dublin, 182
Dulles, Allen, 14–15, 40
Dulles, John Foster, 12, 15, 57, 60
Dutton, Frederick G., 124

Earth, xv–xvi, 131
  burgeoning world economy of,
    xv, 169
  challenge of sustainable
    development on, xv–xvii,
    160–65
  ecological stress on, 164
  as home of our common

humanity, xv–xvi, 67, 83, 158,
  175, 185
  overcrowding on, xv–xvi, 160,
    164
East Berlin, xii, 19, 97, 99, 105, 108
  exodus of Germans from,
    23–24
  see also Berlin Wall
Egypt, Israeli peace accord with,
  150
Eisenhower, Dwight D., 9, 25, 64,
  92, 106
  as supreme Allied commander,
    57
  calls for peace of, 57–60, 61,
    62, 66
  Churchill and, 60–61
  Cold War rhetoric of, 61
  Cuban policy of, 15
  Farewell Address of, 53, 61–62
  hardline advisers of, 57, 60, 62
  JFK's relationship with, 15–16
  "military-industrial complex"
    warning of, 62
  "missile gap" charged to, 12
  nuclear policy of, 13–14, 20–21,
    61, 138
  U-2 spy plane responsibility of,
    18, 61
  see also "Atoms for Peace"
    speech; "Chance for Peace"
    speech
elections, U.S.:
  of 1946, 134
  of 1956, 134*n*
  of 1960, xiii, 9, 10, 12–13, 17, 51,
    52, 134
  of 1962, 28, 35

elections, U.S. (*cont'd*)
  of 1964, 35
  of 1968, 104
  of 1980, 144
England, JFK's 1963 trip to, 96,
  104
environmental change, 160
environmental conservation, 131,
  162, 207
environmental pollution, 131, 162,
  206
  from greenhouse gases, xvi
  industrial, 131, 207
  from radioactive fallout, xvi,
    65, 85, 105–6, 116, 131, 171,
    194–95, 197, 207
Erhard, Ludwig, 47
Etzioni, Amitai, 10–11
Evening Standard (London), 94
Evers, Medgar, assassination of, 86*n*

famine, 164, 183
Farewell Address (Eisenhower),
  53, 61–62
Federal Bureau of Investigation
  (FBI), 40
Fisher, Adrian, 111
Florida, 14, 28
Ford, Gerald R., 143
Foster, William C., *31*, 72, 111
France, 19, 96, 116
  nuclear aspirations and
    acquisition of weapons, 8, 37,
    93, 107, 127, 140, 196
  U.S. relations with, 46, 127
Frankfurt, 97
Fulbright, William, 122
Fulton, Mo., 53, 54

Gallup polls, 126
Galway, 187
Gandhi, Mohandas "Mahatma,"
  118
Geneva, 162, 177, 206
geopolitics, 3, 4, 67
German Democratic Republic
  (East Germany), 19, 22,
    23–25, 203
  Berlin Wall erected by, 24, 27,
    40, 97, 99, 187
  exodus of Germans from,
    23–24
  U.S. nonrecognition of, 125
Germany, Federal Republic of
  (West Germany), 19, 134, 203
  economic and military buildup
    of, 20, 21, 23, 47
  JFK's 1963 visit to, 96–101
  nuclear aspirations of, 21, 23, 37,
    47, 101, 142
  Soviet concerns about, 19–25,
    27, 28, 32, 33, 47, 54, 101, 142
  U.S. relations with, 47, 100–101,
    119, 142
  see also East Berlin; West Berlin
Germany, Nazi, 100
  military aggression of, 5, 47
  U.S. bombing of, 4
  U.S.-Soviet settlement plan for,
    19–20
  see also World War II
Germany, reunification of, 19, 24,
  59, 142
Gilpatric, Roswell, 27–28, 30*n*, *31*,
  39n
globalization, 160
global warming, 161

God, 53, 87, 169, 185
Goldwater, Barry, 35, 123
Gorbachev, Mikhail, 138, 146, 150, 151, 152
Graham, Thomas, 140–41
Greece, 56
greenhouse gases, xvi
Gromyko, Andrei, 29, 109
Guardian (Manchester), 94
Guns of August, The (Tuchman), 42–43

Hard Way to Peace, The (Etzioni), 10
Harriman, Averell, 39, 40, 73*n*, 95
  in test ban negotiations, 109–11, 134*n*, 192–93
Harris poll, 126
Harvard University, 113
Havel, Vaclav, 151
Hawaii, 72, 73, 86
Helsinki Declaration (1975), 151
Hiroshima, 125
  U.S. nuclear bombing of, 4
Hitler, Adolf, 19
  appeasement of, 5–6, 9, 74
  defeat of, 4
  Germany rearmed by, 5
  military aggression of, 5, 80–81
  rise of, 163
Hoover, J. Edgar, 40
House of Commons, British, 51, 82
Hughes, Emmet, 57*n*
human rights, 87–89, 132, 151, 207–8
Hungarian revolution, 187
hunger, 131, 163, 183, 206

Huntington, Samuel, 164
hydrogen bomb, *see* thermonuclear bomb

illiteracy, 131
inaugural address of 1961 (Kennedy), xiii, 11–12, 63–64
  challenges of Cold War and nuclear weapons noted in, 160
  defense of liberty and democracy articulated in, 11, 45
  four precepts of, 41–45
  negotiation vs. fear counseled by, 63–64
  poverty addressed in, 164–65, 169
  power of human choice conveyed in, 64
  religious reference in, 53, 63, 87
  science commitment in, 11, 42, 64
income and wealth disparities, 160
India, 176
  liberation from colonial rule of, 118
  nuclear testing and arsenal of, 140, 141, 150
  Pakistani conflict with, 141, 167
Indochina, 59
integration, 86–89
  in armed services, 89
  National Guard federalized by JFK in support of, 86*n*, 88
  of University of Alabama, 86–87, 88

intercontinental ballistic missiles
   (ICBMs), 32, 146
Interim Agreement on the
   Limitation of Strategic
   Offensive Arms, 143
intermediate-range missiles, 142
Iran:
   culture and history of, 168
   U.S. conflict with, 167, 168
Iranian Revolution, 150
Ireland:
   JFK's 1963 trip to, 96, 101–4,
      102
   U.K. relations with, 102
Irish Dáil speech (Kennedy),
   101–4, 155, 180–89
Irish Parliament (Dáil), 101, 155
"iron curtain," 54, 56
Islam:
   extremism in, 164
   fighting forces of, 150
   struggle between Christianity
      and, 164
isolationism, 24
Israel:
   Egyptian peace accord with,
      150
   nuclear arsenal of, 141
   Palestinian conflict with, 167
   water supply controlled by,
      167
Italy:
   JFK's 1963 visit to, 104–5
   U.S. missiles placed in, 14, 28
Izvestiya, 94, 95

Japanese Navy, 4
Jervis, Robert, 7–8, 14, 147

Johnson, Lyndon B., 30*n*, *31*, *40*,
   124, 135, 142
   Civil Rights Act signed by, 89
   escalation of the Vietnam War
      by, 149
John XXIII, Pope, 46
   death of, 53, 67–68, 69, 105
   JFK influenced by, 53–54,
      67–69, 71
   Pacem in Terris (Peace on
      Earth) encyclical of, 53–54,
      67–69
John Paul II, Pope, 151
Joint Chiefs of Staff, 32, 72, *120*, 146
   nuclear treaty reservations of,
      121, 125, 134
   nuclear treaty support of, 119,
      120–22, 123
Judaism, 52–53
Jupiter missiles, 14, 33–35

Kahneman, Daniel, 33*n*
Kashmir, 141, 188, 209
Kaysen, Carl, 39*n*, 72–73
Keating, Kenneth, 29
Kefauver, Estes, 134*n*
Kennan, George, *40*, 151, 153–54
   security through containment
      advocated by, 45, 153
Kennedy, Edward M. "Ted," 104
Kennedy, Jacqueline, 135
Kennedy, John F., xi–xvi, 3–6, *52*,
   61, *120*, *128*, 148
   annus mirabilis (October 1962–
      September 1963) of, xiii,
      40–41
   assassination of, 13, 67, 134–35
   cabinet of, 30*n*, *31*, 39, *40*, 72

charm and persuasion of, 90, 91, 96, 112, 114

common fate of humanity emphasized by, 67, 83, 155, 157–58, 159, 210

cooperation and patience urged by, 11, 41–45, 48–49, 64, 85–86, 129, 159, 202, 205

domestic political concerns of, 8, 10, 29, 31–32, 35

early executive inexperience of, xiii, 39–40

education of, 4

enduring legacy of, 156–59

faith in humanity of, 133

first thousand days in administration of, 65

foreign policy of, xiv, 12, 15–19, 31–32, 36–37, 39–49

inauguration of, 11–12, 41, 42, 160

Irish Catholic heritage of, 101, 181–82

leadership of, xv, xvi, 48–49, 50, 70, 78–79, 90, 119, 126, 133, 156–58, 166, 168–69

"missile gap" rhetoric of, 12, 27

mistrust of military by, 16, 36

moon mission of, 157, 158, 159

1960 election of, xiii, 9, 10, 12–13, 17

1963 European trip of, 90, 91, 96–105, 113, 119

non-invasion pledge of, 33, 35, 179

nuclear policy of, 8–14, 66–67

nuclear proliferation as major concern of, 139–40

peace strategy and campaign of, xiii–xvi, 3–6, 8–14, 17, 40–49, 52–53, 56, 60, 63–67, 70–88, 91–111, 112–13, 136, 138, 155, 159, 162, 166, 169, 171–79, 201–2

personal courage of, 50, 63, 94, 159

political campaigns of, 9–10, 12, 51, 52, 134

political criticism of, 29, 31–32, 35, 39, 71

public standing of, 35

Pulitzer Prize of, 50

reelection concerns of, 71

rhetorical skills of, 112, 114, 117, 159

Senate career of, 12, 50, 52, 73, 118–19

six arms control steps of, 139, 145

six precepts of, 41–48

space exploration urged by, 42, 64, 67, 159

as student of history, 3–6, 8–10, 42–44, 71, 78

taking up of generational causes urged by, 156–57, 160–68

use of bully pulpit by, 114

world vision of, xiv, 66–67, 70–71

World War II service of, 4, 16

writing of, 4, 5–6, 50, 52

Kennedy, John F., speeches of, xvi, 50, 63–67, 74, 84, 87, 156–62, 179, 209

practicing of, 51

presidential campaign, 9–10, 51

Kennedy, John F., speeches of
  (*cont'd*)
  role of Ted Sorensen in, xiv–xv,
    51–53, 64, 70–73, 169
  use of *antimetabole* in, 64, 65
  see also specific speeches
Kennedy, Joseph P.:
  appeasement of Hitler
    defended by, 5
  as U.S. ambassador to U.K.,
    4, 5
Kennedy, Joseph P., Jr., 4
Kennedy, Kathleen "Kick," 104
Kennedy, Robert F., 36
  assassination of, 104
  as attorney general, 30
  as JFK's adviser, 30, 32, 33, 34
  1968 presidential campaign of,
    104
Keynes, John Maynard, 152, 153
Khrushchev, Nikita Sergeevich,
    xii–xiii, 68, 105
  boasting and brashness of, 28
  correspondence of JFK and, 13,
    17–19, 32–33, 34–35, 37–38,
    44–45, 46–47, 49, 106
  Cuban Missile Crisis and, 14,
    16, 29–38, 46, 48, 135
  domestic opposition to, 28,
    35, 60
  Jacqueline Kennedy's message
    to, 135
  JFK as adversary and partner
    of, xii–xiii, 6, 9–10, 12–13,
    17–19, 21–24, 22, 26–38,
    44–49, 71–72, 85, 92, 94–95,
    101, 105–10, 119, 133–35, 139,
    142, 166, 178, 196–97

  ouster of, 67, 135–36
  peaceful coexistence concept
    of, 18, 28–29, 37–38, 49, 94
  postwar Germany as concern
    of, 21–24, 27, 28, 32, 33, 47
Killian, James R., 113–14
King, Martin Luther, 87
Kissinger, Henry A., 147, 149n
Kistiakowsky, George, 113–14
Korean War, 57
  call for armistice in, 58
Kremlin, 93, 154
Kuznetsov, V. V., 106n

Laos, 40, 194, 200, 201
Latin America, 35
League of Nations, 50, 71, 96, 112,
    120
LeMay, Curtis E., 120, 121, 123
leukemia, 116, 194
Liddell Hart, B. H., 43, 78, 83
Lippmann, Walter, 92
"Long Telegram" (Kennan), 153
Los Alamos Scientific Laboratory,
    123

Malaya, 59
Malenkov, Georgy, 60
Mali, 163, 165
Mansfield, Mike, 120
Mao Zedong, 108
Marshall, George, 163
Marshall Plan, 163, 165
  Soviet rejection of, 56–57
Masefield, John, 74, 171
Massachusetts Institute of
    Technology (MIT), 113
mass migration, 160

Mastny, Vojtech, 133–34
McCone, John, 30*n*, *31*, 121
Macmillan, Harold, 101*n*
  in test ban treaty negotiations,
    44–45, 85, 93–94, 104, 107,
    119, 178, 192
McNamara, Robert, 30*n*, *31*, 32,
    39, 72, 110
medical research, 131, 207
Middle East, 149, 167, 176, 209
  potential use of nuclear
    weapons in, 161
military-industrial complex:
  Soviet, 147, 166
  U.S., 8, 62, 147, 166
missile defense initiative, 144
missiles, 14, 28, 29, 33–34
  anti-ballistic, 143, 145, 146
  ballistic, 12, 32, 143, 146, 204
  intercontinental, 32, 46
  intermediate range, 142
  launchers of, 143
  submarine-based, 146, 204
Molotov, Vyacheslav, 60
moon, 44
  JFK's case for going to, 157, 158,
    159
  possible joint expedition to, 205
  race to, 205
Moscow, 85, 91, 92, 94, 138, 147,
    177
  test ban negotiations and
    signing in, 105–11, 113, 114,
    120, 178, 191, 192
mujahideen, 150
Multilateral Force (MLF), 142
  sharing of nuclear weapons
    with allies in, 21, 28, 29, 37

multiple independently targeted
    warheads (MIRVs), 148
Munich Conference of 1938, 5, 6,
    9, 55–56, 71, 74

Nagasaki, 125
  U.S. nuclear bombing of, 4
Naples, 105
Nationalist China, 57
national liberation movements,
    95, 142
National Security Council, 30
  Executive Committee
    (ExComm) of, 30, *31*, 32–33
National Service Corps, 178
Navy, U.S., 4, 8
Nebraska, 52
Nevada nuclear testing site, 146
New Delhi, 140
New Strategic Arms Reduction
    Treaty (New START), 148
New York Times, 23, 92
Nitze, Paul, 30*n*, *31*, 39
Nixon, Richard M., 143
  Brezhnev's relations with,
    142–43
  China visit of, 142–43
  1960 presidential campaign of, 12
nonaggression agreement, 109, 192
Non-Proliferation Treaty (NPT),
    139–41
  criticism of, 140
  extension of, 140
  inspection site obligations in,
    141
  nuclear-free world aspirations
    of, 141
  partial success of, 139–40

Non-Proliferation Treaty (NPT),
  (*cont'd*)
  signing of, 140
  undermining of, 141
Norris, George, 52
North Atlantic Treaty
  Organization (NATO), 21,
  23, 34, 37, 79, 105, 134, 142
North Korea:
  nuclear arsenal of, 141
  relations of South Korea and, 141
nuclear arms, xi–xiii, 160
  delivery systems of, 8, 67, 139,
    143, 148
  first use of, 4
  fissionable materials used in,
    67, 138, 204
  gradual destruction of, 139
  limiting transfers of, 67, 139
  nations in possession of, 141,
    142, 202–5
  non-proliferation of, 159
  peaceful uses of, 125
  potential use of, in regional
    conflicts, 161
  proliferation and spread of, 8,
    13–14, 21, 41, 42, 63, 66, 81, 85,
    116, 138–41, 143–49, *145*, 160,
    193
  radioactive fallout from, xvi,
    65, 85, 105–6, 116, 131, 171,
    194–95, 197
  reduction in numbers of, xii,
    21–22, 59, 66–67, 139, 143, 148
  Soviet possession and testing
    of, xi, xii, xvi, 3–14, 18, 26–38,
    57, 60, 137, 140, 142–49, *145*,
    151, 200

  stockpiling of, 115, 146–47, 191,
    203
  threat of global war with,
    xi–xiii, xvi, 3, 24–25, 29–38,
    114, 160, 169, 190, 193–94,
    195–96, 203
  U.S. possession and testing of,
    xii, xvi, 3–14, 26–28, 29, 33–
    38, 60, 72, 140–49, *145*, 159,
    175–76
  Western European allies'
    sharing of, 21, 23, 28, 37
nuclear arms control, 10, 11, 21–22,
    60, 114–15, 159, 162, 196
  disarmament talks on, 84–85,
    114, 177, 178, 204
  hardline opposition to, 143, 166
  JFK's six steps to, 139, 145
  mutual gains of, 63, 70–71
  mutual verification in, 204
  reduction of tensions through,
    45, 85, 134, 202
  U.S.-Soviet negotiations on,
    38, 39, 84–85, 114, 143, 144,
    145–46, 147–48, 162
nuclear arms race, xi, xii, xvi, 4,
    6–8, 10, 21–22, 27, 39, 60, 66,
    92, 114, 129, 137–38, 143–49,
    145, 168, 193–194, 202, 203
  concept of deterrence in, 75–76,
    134, 175
  costs of, 41, 42, 57*n*, 58, 59, 75,
    81, 82, 147, 171–72, 174
  danger of technological
    breakdown and false alerts
    in, 44, 63, 75
  dogma of "massive retaliation"
    in, 61

limitation of, 116–17, 127–28,
130, 175, 197, 201
as a prisoner's dilemma, 7, 10,
41, 75
U.S. advantage in, 12, 27–28, 146
nuclear reactor design, 161
nuclear scientists, 119, 122, 123
nuclear submarines, 146, 204
nuclear test ban treaty
(proposed), xvi, 95, 100–101
atmospheric testing issue in, 26,
27, 44, 85, 105, 108, 178, 197, 200
hardline opposition to, xvi, 6, 7,
8, 46–47, 71, 108, 134
JFK's proposal of, 44–45,
66–67, 85
mutual benefits of, 70–71, 82,
117, 130, 142, 175
negotiations on, 21–22, 37, 49,
61, 85, 91, 92, 97, 100–101,
104–11, 114–15, 119, 134,
177–78, 190–93
space tests issue in, 105–6, 108,
111, 114, 191, 197
U.K. involvement in, 44–45, 93,
111, 178, 191, 192
underground tests and onsite
inspections issue in, 61, 95,
105–9, 115, 146, 193, 204
see also Partial Nuclear Test
Ban Treaty
Nunn, Sam, 149n

Obama, Barack, scientific centers
proposed by, 167
O'Brien, Lawrence F., 124
O'Donnell, Kenneth, 24, 30n, 100
"Operation Mongoose," 16

"Operation Niblick," 146
"Operation Wheaties," 60n
O'Reilly, John Boyle, 102, 185
Organization of American States
(OAS), 33
Ormsby-Gore, David, 36, 104

Pacem in Terris (Peace on Earth)
encyclical, 53–54, 67–69
Pacific Ocean, 4
Pakistan:
Indian conflict with, 141, 167
nuclear arsenal of, 140
Palestinians, Israeli conflict with,
167
Paris Peace Treaty, see Treaty of
Versailles
Partial Nuclear Test Ban Treaty
(PTBT), 107–11, 112–34, 149,
204
adoption of, 134
atmospheric tests banned
in, 107–8, 111, 114, 116, 121,
127–28, 146, 191, 192, 194–95,
197, 201
confirmation of, 112–33
domestic politics of, 119
era of détente inaugurated by,
138, 141–43
as first step to peace, 115–16,
118, 129, 136, 199
historic importance of, 126,
133–34, 136–39, 193
hurdles overcome in, 105–11
mutual gains of, 142
opinion surveys on, 125–26
possible violations of, 117,
196–97

Partial Nuclear Test Ban Treaty
(PTBT) (cont'd)
    progress made in wake of,
        138–41
    public support sought for,
        113–18, 119, 125–26
    safeguard provisions in, 121,
        122, 125, 146
    Senate debate and ratification
        of, 91, 111, 112–13, 117–20,
        122–26, 134, 201
    signing of, 111, 113, 119–20, 126,
        127, 132, 134, 145, 201, 210
    space tests banned in, 191, 197
    support of Joint Chiefs of Staff
        for, 119, 120–22, 125
    support of world leaders sought
        for, 127–33
    U.K. involvement in, 44–45, 85,
        93–94, 104, 107, 111, 119, 127,
        128, 134, 191, 192, 201
    withdrawal clause in, 110, 111,
        115, 197
    see also nuclear test ban treaty
        (proposed)
Partial Nuclear Test Ban Treaty
    Speech to the Nation of 1963
    (Kennedy), 113, 114–18,
    190–99
Paul VI, Pope, 105
Peace Corps, 164, 178, 206
Peace Speech of 1963 (Kennedy),
    xiv–xvi, 3, 53, 54, 56, 67, 69,
    70–88, 73, 91–96, 100, 103,
    128–29, 132, 134, 155, 159,
    170–79
Perry, William J., 149n
plague, 131, 206

Plymouth Bay Colony, 159
Poland, communist rule of, 19
Politburo, 29, 106
political right, 31–32, 35, 39–40, 45
Potsdam Conference of 1945, 19
poverty, 59, 163–65, 168
    abolition of, 169
    causes of, 164
    violence and, 163–64
Powers, David, 30n, 100
Powers, Francis Gary, 18, 61
Pravda, 60, 94, 95
prisoner's dilemma, 7, 10, 41,
    66, 75
Profiles in Courage (Kennedy),
    50, 52
Protestantism, 207

racism, 86–89, 207–8
    segregation as, 86n, 89
    tension and violence in, 86n
Reagan, Ronald:
    anti-Soviet rhetoric of, 144
    arms buildup of, 141, 144
    missile defense initiative ("Star
        Wars") of, 144
    "tear down this Wall" speech
        of, 99
    vow to restore U.S. power by,
        144, 151
Reeves, Richard, 51
Reith Lectures of 2007 (Sachs), xiv
Republican Party, 10, 119, 120
    right wing of, 31–32, 35
resource scarcity, 160
Richter, James, 48
Rio Earth Summit of 1992, 162
Rome, JFK's 1963 trip to, 96, 105

Roosevelt, Franklin D., 17, 95
Rusk, Dean, 30*n*, *31*, 39, *40*, 72,
    110, 120, 121, 122
Russia, 55, 61
    economic chaos in, 163
    lawlessness and corruption in,
        148
Russian Revolution, 152

Sahel desert, 164
Sakharov, Andrei, 151
Samuelson, Paul, 166
Saturday Review, 43, 46
Schlesinger, Arthur, Jr., 95, 96–97
Seaborg, Glenn T., 72, 101*n*, 110,
    121, 122–23, 125*n*
security dilemma, 7, 14, 147
Senate, U.S., 71, 96, 108, 111
    Armed Services Committee of,
        124
    Foreign Relations Committee
        of, 122–24
    JFK's career in, 12, 50, 52, 73,
        118–19
    partial test ban treaty debate in,
        117, 120, 122–26, 127, 198
    partial test ban treaty
        ratification by, 91, 111, 112–13,
        118–20, 126, 201
    SALT I ratified by, 143
Senegal, 164
Shaw, George Bernard, 103–4, 186
Shoup, David M., *120*
Shriver, Eunice Kennedy, 101
Shultz, George P., 149*n*
Silvestri, Vito N., 97*n*
"Sinews of Peace" speech
    (Churchill), 53–56

Smith, Jean Kennedy, 101
socialism, 35
Solzhenitsyn, Alexander, 151
Somalia, 165
    U.S. military action in, 163
Sorensen, Gillian, xiv
Sorensen, Theodore C. "Ted,"
    xiv–xv, 52, 93, 105, 124–25
    background and youth of,
        52–53
    insight and eloquence of, 51
    intellect and moral force of, xiv,
        51–52
    as JFK's counselor and
        speechwriter, xiv–xv, 26, 30*n*,
        31, 33, 51–53, 64, 70–73, 110,
        169
Sorensen, Thomas, 93
"Sources of Soviet Conduct, The"
    (Kennan), 153
South Africa, 140, 203
South America, 149
South Korea, 140
    relations of North Korea and,
        141
Soviet Air Force, xi
Soviet Communist Party, xii, 3, 95,
    153
Soviet Council of Ministers, 106
Soviet Union:
    Afghanistan invaded by, 143,
        150, 151
    Chinese relations with, 93, 95,
        107, 108, 127, 142, 194
    collapse of, 148, 152
    concerns regarding West
        Germany, 19–25, 27, 28, 32,
        33, 47, 54, 101, 142

Soviet Union (*cont'd*)
    Cuban alliance with, 14–18, 27,
        28–38, 134
    cultural exchanges between
        U.S. and, 134
    Eastern European satellites of,
        20, 59, 92, 151
    economic crisis in, 142, 148, 152
    as "evil" empire, 144
    "hotline" communication
        between U.S. and, 134, 204
    internal power struggle in, 60
    "iron curtain" references to,
        54, 56
    limited ballistic missile capacity
        of, 12
    military aggression of, 142, 194
    military-industrial complex in,
        147, 166
    morale in, 142, 147, 150
    nuclear arsenal and testing of,
        xi, xii, xvi, 3–14, 18, 26–38, 57,
        60, 137, 140, 142–49, *145*, 151,
        200
    propaganda of, 173–74
    rapid industrialization of, 152
    repression in, 142
    scientific and cultural
        achievements of, 174
    space program of, 27
    technological isolation of, 152
    U.K. relations with, 80, 85
    U.S. relations with, xi–xiii, xvi,
        3–25, 26–38, 39–40, 42–49,
        57–65, 70–72, 74–86, 87–88,
        91–97, 111, 133–35, 141–53,
        166, 168, 172–78, 190–98,
        202–5

    World War II alliance of U.S.
        and, 17, 19, 80, 82
    World War II casualties of, 19,
        80–81, 174
space:
    issue of nuclear testing in,
        105–6, 108, 111, 114, 139, 197
    JFK's call for exploration of, 42,
        64, 67, 159
    Soviet program in, 27
    U.S. and Soviet cooperation in,
        134, 205
    U.S. and Soviet race to, 159
Stahr, Elvis, Jr., 42–43
Stalin, Joseph, 19–20, 136
    brutality and paranoia of, 57
    death of, 53, 57, 60*n*, 61
State Department, U.S., 40, 110,
    122, 124
Stevenson, Adlai, 26–27
    as UN ambassador, 27, 134*n*
Strategic Arms Limitation Talks
    (SALT I), 143
Strategic Arms Limitation Talks
    (SALT II), 143
Strategic Arms Reduction Treaty
    (START I), 148
Strategic Arms Reduction Treaty
    (START II), 148
Supreme Court, U.S., civil rights
    rulings in, 86*n*
sustainable development, xv–xvi,
    160, 161–65
    defining goals of, 161–62
    economic growth in, 161–62,
        164
    peace through, 163–65
    practical work of, 167

Taiwan, 140

Taylor, Maxwell, 30*n*, *31*, 72, 120–22, *120*, 123

technological change, 160

Teller, Edward, 123

thermonuclear bomb, 4, 38, 116, 123, 137–38, 195, 204

Soviet testing of, 26

Thomas, Evan, 57*n*

Thompson, Llewellyn, 30*n*, *31*, *40*, 72

Times (London), 94

Todd, Emmanuel, 230*n*

Trachtenberg, Marc, 21

Treaty of Versailles, 5

Truman, Harry, 54

Truman Doctrine, 56

Tsar Bomba, 26

Tuchman, Barbara, 42–43

Turkey, 56

U.S. nuclear missiles in, 14, 28, 29

withdrawal of U.S. missiles in, 33–34

U-2 spy planes, 12

Cuban surveillance by, 29, 30, 36

grounding of, xi, 44

international law violated by, 18*n*

shooting down of, 18, 36, 61

Unitarian Church, 52–53

United Kingdom, 4–6, 19, 116, 196

Irish relations with, 102, 183, 185

nuclear arsenal of, 8, 37, 140

press of, 93–94

Soviet relations with, 80, 85

special relationship of U.S. and, 54–55, 93–94, 104

test ban treaty participation of, 44–45, 85, 93–94, 104, 107, 111, 119, 127, 128, 134, 178

see also England

United Nations (UN), xiv, 27, 58, 91, 176, 187, 188, 191, 200, 201, 206–10

Charter of, 55, 56, 205, 207, 210

General Assembly of, 53, 60, 65–67, 84, 127–33, *128*, 134, 167, 188, 200–201

permanent members of Security Council of, 140

United Nations Speech of 1961 (Kennedy), 53, 65–67, 71, 84–85, 127, 130, 132–33, 139, 200, 210

United Nations Speech of 1963 (Kennedy), 127–33, *128*, 134, 167, 200–210

United States, 132

Chinese relations with, 142–43

foreign aid policy of, 165, 206

French relations with, 46, 127

"hotline" communication between Soviet Union and, 134, 204

Irish relations with, 182–83, 188

military action in destabilized countries by, 163–64

military-industrial complex in, 8, 62, 147, 166

nuclear arsenal and use of, xii, xvi, 3–14, 26–28, 29, 33–38, 60, 72, 140–49, *145*, 159, 175–76

United States (*cont'd*)
political right in, 31–32, 35,
39–40, 45
Soviet cultural exchanges with,
134
Soviet relations with, xi–xiii,
xvi, 3–25, 26–38, 39–40, 42–
49, 57–65, 70–72, 74–86, 87–
88, 91–97, 133–35, 141–53, 166,
168, 172–78, 190–98, 202–5
U.K. relations with, 54–55,
93–94, 104
values and traditions of, 154
West German relations with,
47, 100–101, 119, 142
U.S. Information Agency (USIA), 93

Versailles, Treaty of, 5
Vienna Summit of 1961, 19, 22–23,
22, 24, 26, 27, 40
Vietnam War, 40, 67, 89
escalation of, 149
U.S. losses and costs in, 150
Vietnamese losses in, 149
Voice of America, 94

Walesa, Lech, 151
Wallace, George, 86*n*
Wall Street Journal, 149*n*
Washington, D.C., 105, 117, 177,
182, 198
Washington, George, 182
farewell address of, 61–62
Washington Post, 92
weather research rockets, 44
weather satellites, 131, 207
West Berlin, 19, 22, 25, 89, 200,
201

JFK in, 96–101, 187
JFK's *"Ich bin ein Berliner"*
speech in, 97–101, *98*, 157
prosperity of, 99
U.S. commitment to, 176
Western access to, 22, 23, 45,
176
Westminster College, 53
Wheeler, Earle G., *120*
White House, 46, 52, 61, 73, 88,
124
architectural design of, 182
tape recording in, 30
Why England Slept (Kennedy),
5–6
Wilson, Donald, *31*
Wilson, Harold, 94–95
Wilson, Woodrow, 74, 118, 121,
124, 170
academic career of, 171
incapacitating stroke of, 96
League of Nations failure of, 50,
71, 112, 120
World Crisis, The (Churchill),
4, 42
World Health Organization, 131,
207
World War I, 3–4, 5, 14, 47, 160
hope for lasting peace after, 96
misjudgments and false
premises at the root of, 42
opposition to U.S. entry into,
52
World War II, 3–6, 14, 47, 160
battles of, 51
bombs dropped in, xii, 4, 75,
116, 171
D-Day invasion in, 104

Eisenhower as supreme Allied
    commander in, 57
ending of, 19, 53, 60, 206
JFK's service in, 4, 16
lead-up and beginning of,
    5–6, 51
prisoners of war in, 58

Soviet casualties in, 19, 80–81,
    174
see also Allies (World War II)

Yemen, 165
    U.S. military action in, 163
Yugoslavia, 20

ABOUT THE AUTHOR

JEFFREY D. SACHS is a world-renowned professor of economics, leader in sustainable development, senior UN advisor, bestselling author, and syndicated columnist whose monthly newspaper columns appear in more than one hundred countries. He is the director of the Earth Institute at Columbia University and Special Advisor to UN secretary-general Ban Ki-moon on the Millennium Development Goals, which are designed to reduce extreme poverty, disease, and hunger, a position he also held under former UN secretary-general Kofi Annan. Sachs directs the UN Sustainable Development Solutions Network on behalf of the secretary-general. He has received many honors around the world, including the Sargent Shriver Award for Equal Justice, India's Padma Bhushan award, Poland's Commanders Cross of the Order of Merit, and many honorary degrees. He has twice been named among the hundred most influential leaders in the world by *Time* magazine.

ABOUT THE TYPE

This book was set in Minion, a 1990 Adobe Originals typeface by Robert Slimbach. Minion is inspired by classical, old-style type-faces of the late Renaissance, a period of elegant, beautiful, and highly readable type designs. Created primarily for text setting, Minion combines the aesthetic and functional qualities that make text type highly readable with the versatility of digital technology.